BRIDGE VISION
PRODUCTION, LLC

Title: Organic-Way Mathematics: The Essence of Problem Solving - Grades 5-6-7 - Teacher Book
Author: Archangelo Joseph, Ed.D.
Layout and Cover: Kate Keverline

Bridgevision Production LLC
PO BOX 9767
Washington, DC 20016
erilandesully@gmail.com
www.bridgevisionllc.com

Library of Congress Catalogue-in-Publication Data is available upon request.
Originally published: 2023.

ISBN: 978-1737678984

TABLE OF CONTENTS

TABLE OF CONTENTS
(continued)

PREFACE

Organic-way Mathematics: The Essence of Problem Solving (OMEPS) is the flagship of Organic-way Mathematics Education. Underpinned by an 8-year doctoral study (Joseph 2020) under the aegis of Concordia University Chicago, OMEPS has captivated teachers and students through several years of informal piloting at public schools in Brooklyn, New York as of 2007 and in the Artibonite Department of Haiti, as of 2017, particularly in Saint-Marc and in Gonaïves.

Organic-way Mathematics: The Essence of Problem Solving is based on the developmental constructivism of Jean Piaget blended with the socio-cultural theory of Lev Vygotsky. The confluence of the two constructivist philosophies frames the tenets of our Organic-way education project, which includes Organic-way Mathematics: Profound Conceptual Understanding (MAPCU).

In opposition to prominent behaviorists such as John Broadus (J.B.) Watson, Edward Thorndike, and Burrhus Frederic (B.F.) Skinner, constructivists believe in the students' agency. That is, the students must play an active part in the learning process, as evidenced by these two quotes:

"To understand is to discover, or to reconstruct by discovery, and such conditions must be complied with, if in the future, individuals are formed, who are capable of production, and not repetition" (Piaget 1973).

"Children construct their knowledge. Development cannot be separated from its social context. Learning can lead development. Language plays a central role in mental development. Students learn best in company of others, or the more able persons" (Vygotsky 1978).

In practice, the constructivist epistemology engenders constructed-response questions, requiring justifications of answers in contrast to multiple-choice questions. In fact, constructed-response questions compel students to compose their own responses from scratch. On the other hand, despite its limited benefits and popularity among students, multiple-choice questions (MCQ) are proven less effective at leveraging deeper conceptual understanding. For the MCQs leave too much room for students to guess the correct answer, prompting the creation of OMEPS.

Organic-way Mathematics: The Essence of Problem Solving (OMEPS) has come to serve as a supplement to any math program. Based on Joseph (2020), the traditional math programs are insufficient to leverage student profound understanding of essential concepts. To fill in the gaps, OMEPs has come to suggest a way for the teacher to create space for a community of learners, including herself, to "do" mathematics, negotiate meanings, and arrive at the most efficient solution. It is worth noting that OMEPS does not introduce mathematical concepts. Rather, it presumes that the concepts embedded in the OMEPS problem have been already covered.

In sum, on purpose OMEPS Grades 5-6-7 focuses on a troublesome group of foundational and fundamental concepts (e.g., fraction, percentage, decimal, ratio, and proportion). These concepts are arguably the most troubling to upper-grade students. The lack of mastery of these topics created

some profound gaps that are difficult to be filled in later grades. For example, whereas a high-school student may tackle with ease the function $f(x) = 5x + 6$, she may be mystified by the function $f(x) = 1/2\ x + 5x + 6$ due to the presence of the fraction ½ as the slope. That's why the main intent of OMEPS is to demystify the nature of fractions, among the others, through real-life application, continuous reinforcement, and standards-based assessment.

Below are underscored the key characteristics of a typical 90-minute OMEPS lesson.

Characteristics of OMEPS Grades 5, 6, and 7

- Project-based and student-centered learning.
- Real-life application, reinforcement, and assessment most foundational and fundamental concepts —The Gang of Five—Fraction, Decimal, Percentage, Ratio, & Proportion.
- Prototype problems or tasks with differing solution strategies.
- Emphasis on annotated reading of the mathematical story problems.
- Logical reasoning, rigor, and flexibility in alignment with any state curriculum and standard.
- Hands-on approach with realia (household items and environmental materials).
- Usage of multiple intelligences (visual, auditory, kinesthetic, etc.)
- Emphasis on small-group cooperative learning and whole-class discussions, and reflection.
- One-to-one support for low-performing students

Below is a striking testimonial highlighting the potency of OMEPS.

"Organic Math [Organic-way Math] *is an instructional resource that closely follows Piaget's constructivist theory. At the beginning of each problem, students find themselves in a state of disequilibrium, thus forcing them to activate their prior knowledge. Organic Math encourages the exploration of problem solving. Each student actively participates in trying to solve the task in his or her unique way. It encompasses cooperative group work, spiral review of previous math skills while still introducing new concepts. The premise is to teach / re-teach multiple math skills that are embedded in the word problem; skills that are not taught in isolation. Organic Math is a flexible tool that can be utilized in the classroom at the teacher's discretion. It aids in the preparation of State Assessments because it is aligned with the State Standards. Following are some key points of Organic Mathematics:*

- *Flexibility in using the program in class.*
- *Spiral Review: All skills are taught in conjunction of one another.*
- *Kids are allowed to be in disequilibrium, and it is acceptable.*
- *Teachers are truly only a facilitator. Through accountable discourse, students find the solutions or approximate them.*

- *Word problems are interesting and connected with ELA (English Language Arts, i.e., reading and writing)*
- *Flexibility and Test Prep.*

I conclude by saying that I would choose Organic Math over many other programs, some of which we are currently using officially.

Irma Mendez, 6C Classroom Teacher
The Bilingual School/PS 189K, Brooklyn, New York
June 17, 2007

PROBING MATHEMATICAL QUESTIONS

The landmark Equity Educational Opportunity Study (EEOS), Coleman (1996) finds that teachers are the most important resources that students ever had in school—not the books, not the principal, not the curriculum. Many frameworks (Bloom's Taxonomy and Depth of Knowledge) would agree that leveraging student understanding depends on the appropriateness of the questions that teachers pose. Developed by Kersaint et al (2013) in *Orchestrating Mathematical Discourse to Enhance Student Learning*, following are sets of questions that teachers could posed to improve student learning.

To help students work together to make sense of mathematics.

1. *What strategy did you use?*
2. *Do you agree?*
3. *Do you disagree?*
4. *Would you ask the rest of the class that question?*
5. *Could you share your method with the class?*
6. *Can you convince us that your ideas make sense?*
7. *What do you think about what [Paul] said?*
8. *Can you restate what [Paul] said?*
9. *Would any one like to add to this?*
10. *Have you discussed your ideas with the group?*
11. *Did anyone get a different answer?*
12. *Did anyone in the group have any reservations?*
13. *Did anyone get a fair chance to share ideas?*
14. *How would you explain this to [Paul]? who missed class today?*
15. *Refer questions raised by students back to class.*

To help students rely on themselves to find if something is mathematically correct.

1. *Is this a reasonable answer?*
2. *Does that make sense?*
3. *Why do you think that? Why is that true?*
4. *Can you draw a picture or a model to show that?*
5. *How did you reach that conclusion?*
6. *Does anyone want to revise [Paul's]?answer?*
7. *How were you sure that your answer is correct?*

To help students learn to reason mathematically.

1. *How did you begin to think about this problem?*
2. *What is another way you could solve this problem?*
3. *How could you prove that?*
4. *Can you explain how your answer is different from or the same as [Paul's]?*
5. *Let's see if we can break it down. What would the parts be?*
6. *Can you explain this part more specifically?*
7. *Does that always work?*
8. *Is that true for all cases?*
9. *How did you organize your information? Your thinking?*

To help students evaluate their own processes and engage in productive peer interaction.

1. *What do you need to do next?*
2. *What have you accomplished?*
3. *What are your strengths and weaknesses?*
4. *Was your group participation appropriate and helpful?*
5. *Dis you help other students with problem comprehension?*
6. *What is this problem about? What can you tell me about it?*
7. *Do you need to define or set limits for the problem?*
8. *How would you interpret that?*
9. *Would you please reword that in simpler terms?*
10. *Is there something that can be eliminated or that is missing?*
11. *Would you please explain that in your own words?*
12. *What assumptions do you have to make?*
13. *What do you know about this part?*
14. *Which words were most important? Why?*

To help students learn to conjecture, invent and solve problems.

1. *What would happen if ___?*
2. *Do you see a pattern?*
3. *What are some possibilities here?*
4. *Where would you find the information that you need?*
5. *How would you check your seps or your answer?*
6. *What did not work?*
7. *How is your solution method the same as or different from [Paul's]?*
8. *Other than retracing your steps, how can you determine the accuracy of your answers?*
9. *What decision do you think he or she should make?*
10. *How did you organize the information? Do you have a record?*
11. *How could you solve this using (tables, trees, lists, diagrams, etc.)?*
12. *What have you tried? What steps did you take?*
13. *How would it look if you used these materials?*
14. *How would you draw a diagram or make a sketch to solve the problem?*
15. *Is there another possible answer? If so, explain.*
16. *How would you research that?*
17. *Is there anything you've overlooked?*

18. *How did you think about the problem?*
19. *What was your estimate or prediction?*
20. *How confident are you in your answer?*
21. *What else would you like to know?*
22. *What do you think comes next?*
23. *Is the solution reasonable, considering the context?*
24. *Did you have a strategy? Explain it.*
25. *Did you have a design? Explain it.*

To help students learn to connect mathematics, its ideas and its application.

1. *What is the relationship of this to that?*
2. *Have we ever solved a problem like this before?*
3. *What uses of mathematics did you find in the newspaper last night?*
4. *What is the same?*
5. *What is different?*
6. *Did you use skills or build on concepts that were not necessarily mathematical?*
7. *Which skills or concepts did you use?*
8. *What ideas have we explored before that were useful in solving this problem?*
9. *Is there a pattern?*
10. *Where else would this strategy be useful?*
11. *How does this relate to ___?*
12. *Is there a general rule?*
13. *Is there a real-life situation where this could be used?*
14. *How would your method work with other problems?*
15. *What other problem does this seem to lead to?*

To help students persevere

1. *Have you tried making a guess?*
2. *What else have you tried?*
3. *Would another recording method work as well or better?*
4. *Is there another way to (draw, explain, say) that?*
5. *Give me another related problem. Is there an easier problem?*
6. *How would you explain what you know right now?*

To help students focus on the mathematics from activities.

1. *What was one thing you learned (or two, or more)?*
2. *Where would this problem fit on our mathematics chart?*
3. *How many kinds of mathematics were used in this investigation?*
4. *What were the mathematical ideas in this problem?*
5. *What is the mathematically different about these two situations?*
6. *What are the variables in this problem? What stays constant?*

THE FIVE MAIN STAGES OF OMEPS

PRE-LESSON
TEACHER PREPARATION

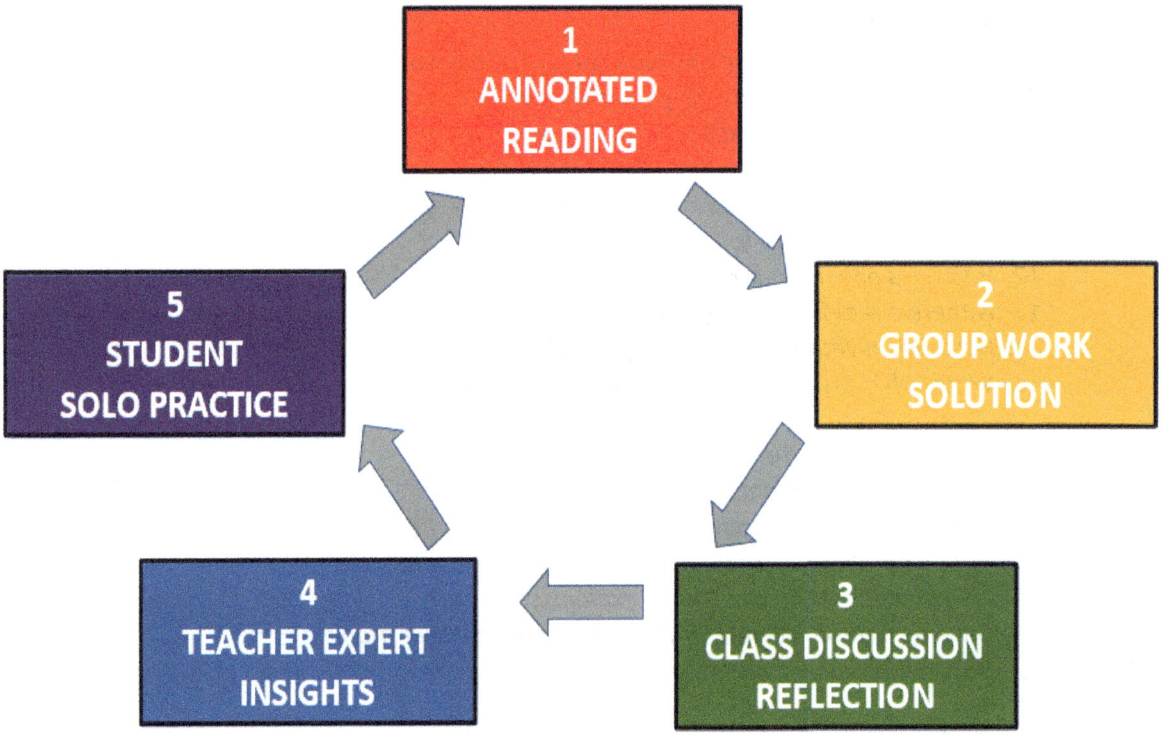

POST-LESSON
TEACHER REFLECTION

STUDENT & TEACHER ROLES & RESPONSIBILITIES

PRE-LESSON TEACHER PREPARATION

Teacher

- develops the lesson plan, anticipating areas of individual student strengths, challenges and potential support;
- prepares a line of thought-provoking/probing questions to leverage student understanding;
- teaches the concepts and skills embedded in the problem.

1

ANNOTATED READING (5 m)

Students

- read the problem several times for profound understanding;
- underline key words and phrases;
- complete the Facts & Questions Table;
- jot down some solution ideas and get ready to join the group.

ANNOTATED READING

Teacher

- ensures that all students are close reading, getting ready for group work.
- Provides reading support to low performers.

2

GROUP WORK SOLUTION (15 min)

Students

- shares understanding of problem and solution strategies;
- select the most efficient strategy;
- educate each other on the group strategy;
- prepare the group solution strategy for presentation
- Take notes on novel ideas.

GROUP WORK SOLUTION

Teacher

- circulates the room to ensure all group dynamics.
- reminds groups to leave no one behind.
- stops the lesson and provides necessary pointers.
- urges all students to take notes of novel ideas;
- Provide minimal scaffolding to groups in needs.
- take notes of group dynamics and learning.

3

CLASS DISCUSSION & REFLECTION (45 min)

Students

- Share the group strategies to the class;
- defend ideas convincingly;
- validate and comment on others' ideas;
- pose questions requesting clarifications;
- take notes on novel ideas;
- Reflect metacognitively on learning.

CLASS DISCUSSION & REFLECTION

Teacher

- signals the beginning of the class discussion;
- select sequence for group presentations;
- ensures the sharing of varied ideas;
- advances the discussion with hypothetical questions;
- takes notes of class dynamics and learning.
- Invites students to share their reflection.

4

TEACHER EXPERT INSIGHTS (10 min)

Students

- listen attentively to teacher's insights;
- pose questions
- inform the teacher on lingering challenges;
- continue to take notes on novel ideas;

TEACHER INSIGHTS

Teacher

- uses observations and notes on class dynamics to inform mini-lessons
- debunk student lingering misconceptions.
- bring to the fore students for their novel ideas;
- summarizes the lesson, underscoring the key points.
- Connects the lesson with the curricula/standards

5

SOLO PRACTICE (15 min)

Students

- use ideas and skills learned from the recent activity
- work individually and quietly;
- complete Solo Practice assessment
- submit Solo Practice assessment;
- return all math materials to the teacher.

SOLO PRACTICE

Teacher

POST-LESSON TEACHER REFLECTION

Teacher

- reminisces all stages of the lessons to bring to life group dynamics, class dynamics, student learning, or the lack thereof;
- examines the notes, observations, and Solo Practice scores to determine the degree of student learning;
- decides on the implications for the subsequent lesson;
- decides whether the concept(s)//skill(s) must to be retaught as a subsequent mini lesson or re-visited in the next problem.

PROBLEM-1
OrgaMath Time

GROUP TASK-1

The 90-minute Organic-way Math lesson allocates time to its components, as follows:

1. Annotated Reading: 1/18 of the total time.
2. Group Solution Annotated Reading: The triple of the
3. Class Discussion: 2,700 seconds
   Teacher Insights: ?
4. Solo Practice: 0.25 hour

- Determine the amount of time (in minutes) left for Teacher Insights.
- Complete the ORGAMATH Time Table.
- Complete the ORGAMATH Time Circle Graph.
- Analyze the data and write your observation.

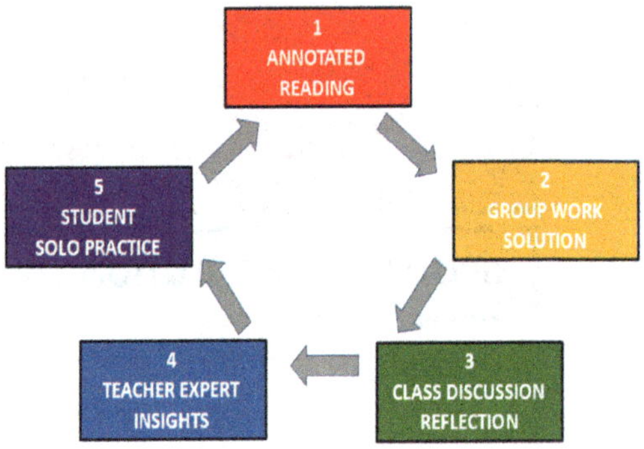

Answer will vary.

<u>Possible solution:</u>
Entire lesson = 90 min.
Annotated Reading = $1/18 \times 90 = 5$ min.
Group Solution = $3 \times 5 = 15$ min.
Class discussion = $2{,}700 \div 60 = 45$ min.
Solo Practice = $0.25 \times 60 = 15$ min
Teacher Insights = $90 - (5 + 15 + 45 + 15) = 10$ min

ORGAMATH Time Table

OMEPS Stage	Allotted Time	Fraction	Notation in Decimal	Percentage
Annotated Reading	5 min.	$\frac{5}{90} = \frac{1}{18}$	0.05	5%
Group Solution	15 min.	$\frac{15}{90} = \frac{1}{6}$	0.166	17%
Class Discussion	45 min.	$\frac{45}{90} = \frac{1}{2}$	0.50 = 0.5	50%
Teacher Insights	10 min.	$\frac{10}{90} = \frac{1}{9}$	0.11	11%
Solo Practice	15 min.	$\frac{15}{90} = \frac{1}{6}$	0.166	17%
Whole Lesson	90 min.	$\frac{90}{90} = 1$	1.00	100%

Observation

We have observed that most time or 50% has been allocated to Class Discussion. It makes us think that this stage of Organic-way Math is deemed the most important in the learning process. Also, it is interesting to see that reading, one modality of language arts, is indeed valued in ORGAMATH, being the first stage and taking 5% of the lesson. Moreover, we have seen that students have to work cooperatively and individually. Each of these stages (Group Solution and Solo Practice) is allocated 15 minutes or 17% of the total time. It is evident that there is no mini lesson at the beginning. But after class discussion, the teacher intervenes probably to wrap up and to debunk some deep-seated students' misconceptions.

ORGAMATH Time Circle Graph

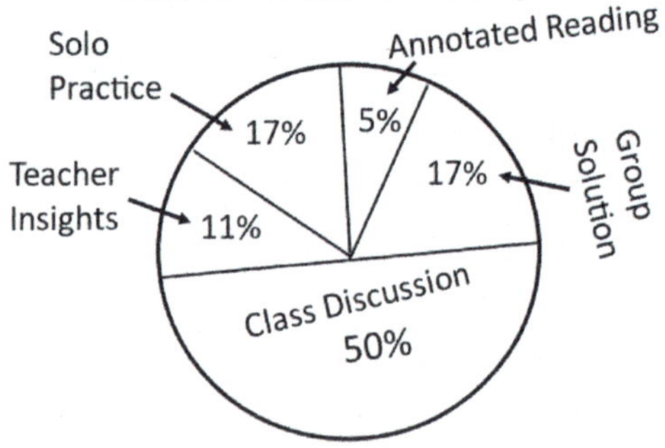

SOLO PRACTICE-1

Imagine that the allotted time for the entire Organic-way Math lesson were one hour. How much time would be proportionally devoted to each lesson stage of the lesson. Complete the relevant table and circle graph.

Answer will vary.

Possible response:
1 hour = 60 minutes.
The Law of Proportion can be used to find out the amount of time allotted to each stage of the 60-minute Organic-way Math lesson.

Annotated Reading:
5/90 = x/60 = 3.33 min.
Group Solution:
15/90 = x/60 = 10 min.
Class Discussion:
45/90 = x/60 = 30 min.
Teacher Insights:
10/90 = x/60 = 6.67 min.
Solo Practice:
15/90 = x/60 = 10 min.

ORGAMATH Time Table

OMEPS Stage	Allotted Time	Fraction	Notation in Decimal	Percentage
Annotated Reading	3.33 min.	$\frac{5}{90}=\frac{1}{18}$	0.05	5%
Group Solution	10 min.	$\frac{15}{90}=\frac{1}{6}$	0.166	17%
Class Discussion	30 min.	$\frac{45}{90}=\frac{1}{2}$	0.50 = 0.5	50%
Teacher Insights	6.67 min.	$\frac{10}{90}=\frac{1}{9}$	0.11	11%
Solo Practice	10 min.	$\frac{15}{90}=\frac{1}{6}$	0.166	17%
Whole Lesson	60 min.	$\frac{90}{90}=1$	1.00	100%

ORGAMATH Time Circle Graph

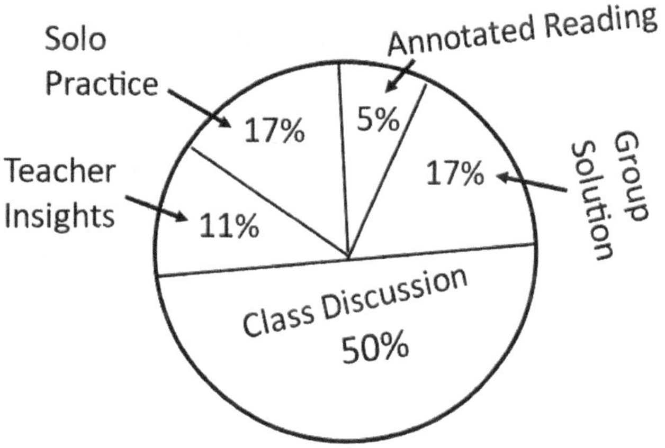

PROBLEM-2
Student-Teacher Ratio

GROUP TASK-2

A lower-grade Co-teaching class requires a student-teacher ratio of 5:1. Two- thirds of the students are boys and the rest are girls. The class has 15 students.

- Determine the number of students of each gender.
- Including teachers. complete the Table Student-Teacher Ratio.
- Complete the Circle Graph for Student-Ratio.

Answer will vary.

<u>Possible Response:</u>
Number of boys: 2/3 ×15 = 10.
Number of girls: 15 – 10 = 5
Number of teachers, from 3:1 ratio
That is, for every 5 groups of students, there are 1 teacher. In 15 there are 3 groups of 5. So, there are 3 teachers working with the class of 15 students.

Table for Student-Teacher Ratio				
People	Number	Fraction	Decimal	Percentage
Boys	10	$\frac{10}{18} = \frac{5}{9}$	0.555	55.5%
Girls	5	$\frac{5}{18}$	0.277	27.8%
Teachers	3	$\frac{3}{18} = \frac{1}{6}$	0.166	16.7%
Total	18	$\frac{18}{18} = 1$	1.00	100%

Circle Graph for Student -Teacher Ratio

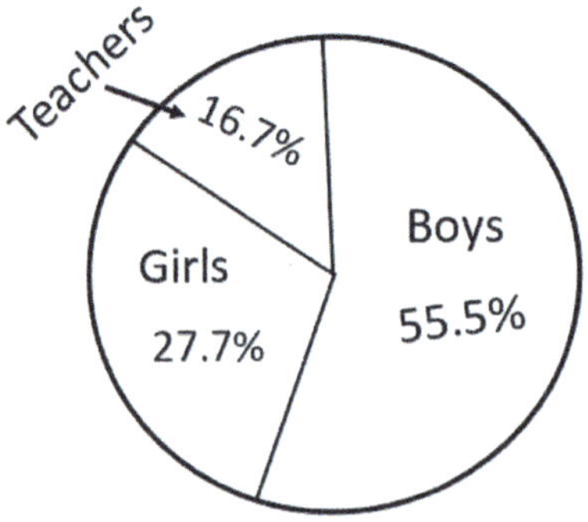

SOLO PRACTICE-2

- Imagine that the same school had 200 students, and all the classes are co-teaching with the same ratio 5:1 of teachers to students. Determine the number of students of each gender.
- Including teachers, complete the Table Student-Teacher Ratio.
- Complete the Circle Graph for Student-Ratio.

Answer will vary.

Possible Response:
All the numbers of people will be computed proportionally to the data on the Group Task, with a ratio 5:1 of students to teachers.
So, the same percentages can be used to compute the numbers of boys, girls, and teachers.
Boys = 55.55 × 240 = 133
Girls = 27.8 × 240 = 67
Teachers = 16.7× 240 = 40

Table for Student-Teacher Ratio

People	Number	Fraction	Decimal	Percentage
Boys	133	$\frac{133}{240}$	0.555	55.5%
Girls	67	$\frac{67}{240}$	0.277	27.8%
Teachers	40	$\frac{40}{240} = \frac{1}{6}$	0.166	16.7%
Total	240	$\frac{240}{240} = 1$	1.00	100%

Circle Graph for Student -Teacher Ratio

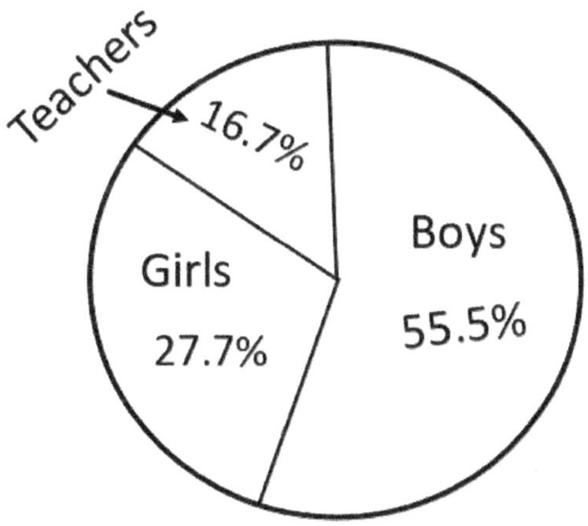

PROBLEM-3
Mixed Beans

GROUP TASK-3

1. Explore your batch of mixed beans.
2. Estimate the total number to be compare later with the actual number.
3. Group them by type or color.
4. Count the number pf bean in each group.
5. Complete the Table for Mixed Beans and Circle Graph for Mixed Beans, assuming that each bean is worth $0.15. Use as many rows as you need.
6. Write your observation.

Student responses will vary and be based on the varied quantity of beans that each group has. Let's suppose we have a total of 20 beans: 10 Black, 3 Green, Pinto 2, and 5 Red.

Table for Mixed Beans

Color	Quantity	Fraction	Decimal	Percentage	Cost
Black	10	$\frac{10}{20}=\frac{1}{2}$	0.50	50%	$1.50
Green	3	$\frac{3}{20}$	0.15	15%	$0.45
Pinto	2	$\frac{2}{20}=\frac{1}{10}$	0.10	10%	$0.30
Red	5	$\frac{5}{20}=\frac{1}{4}$	0.25	25%	$0.75
Total	20	$\frac{20}{20}=1$	1.00	100%	$3.00

Observation

Answers will vary.

<u>Possible Response:</u>

I (we) observe there are more black beans than other colors. In fact, 50% of the beans are black. There are as many as black beans as the other beans combined. But only 10% of the beans are pinto. In between, we have 15% of green beans and 25% of red beans.

Circle Graph for Mixed Beans

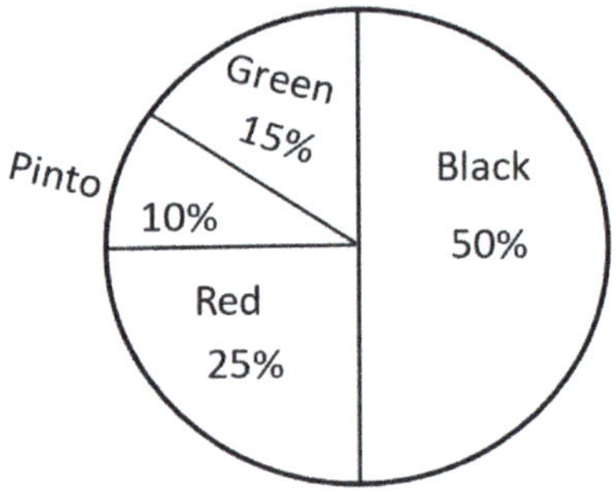

SOLO PRACTICE-3

Student responses will vary and be based on the varied quantity of beans.

PROBLEM-4
Supermarket Shopping

GROUP TASK-4

Based on the accompanying food flyers go to the market and purchase the food items below.
Round all costs to the nearest dollar.
Complete the Table for Supermarket Shopping.
Complete the Circle Graph for Supermarket Shopping.
Write your observation.

- 4 packs of sweet corn
- 3 pints of strawberry
- 5 bags of Cello spinach
- 10 lbs. of chicken drumsticks

Note: Round all costs to the nearest cent. Copy supermarket flyers for student groups.
Suggestion: Next time, make space for students cut the items out and glue them on the table.

Table for Supermarket Shopping

Picture	Quantity	Cost	Fraction	Decimal	Percentage
	4 packs	$4 \times 1.89 =$ $7.56	$\frac{7.56}{23.08}$	0.327	33%
	3 pints	$3 \times 1.29 =$ $3.87	$\frac{3.87}{23.08}$	0.167	17%
	5 bags	$5 \times 0.99 =$ $4.95	$\frac{4.95}{23.08}$	0.214	21%
	10 pounds	$10 \times 0.67 =$ $6.70	$\frac{6.70}{23.08}$	0.290	29%
Total Budget	xxxxxxx	$23.08	$\frac{23.08}{23.08}$	0.998	100%

Observation

(Answer will vary).

<u>Possible Response:</u> I (we) observe the cost ($7.56) of corn is about twice as much as the cost ($3.87) of the strawberries. Also, the cost of the corn is one-third of the budget. Finally, I was having trouble adjusting the precents when doublechecking for the total of 100%.

Circle Graph for Supermarket Shopping

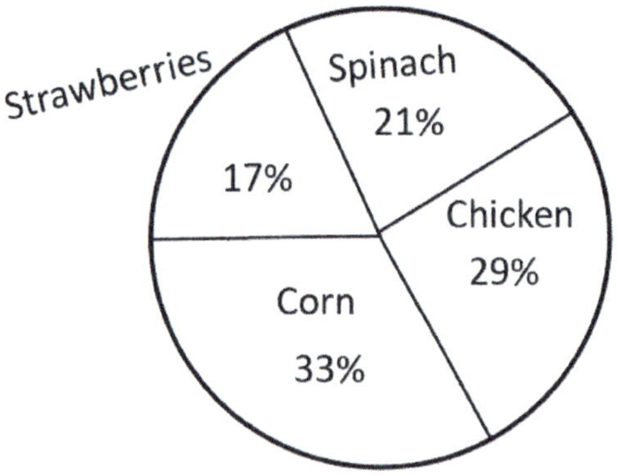

Nicey's Food Mart
AGENTS FOR: MATH POWER | MATH POWER ESL
The fastest way to learn Mathematics
Sweet Corn on Cob
Pkg. of 5
1.89
Cello Spinach
.99 ea
5 lb Bag Florida Oranges
2.99
Fresh Strawberries
By The Pint
1.29 pint
Red or Green Leaf Lettuce
.79 ea
Product of Jamaica
Foska Oats
400 g
Reg. $2.79
1.99
FOSKA OATS
Grace Coconut Water
520 ml
.99
Reg. $1.29
Caribbean Choice or Tropical Estate Ackees
540 ml
4.39
Reg. $5.99
Vitamalt Drink
.89
Reg. $1.49
Geeta Yellow Split Peas
2 kg
.99
Reg. $1.79
Grace Hot & Spicy Vienna Sausage
3/1.99
.99 ea.
Please Turn Over For More Specials
Specials in Effect Thu. April 4, 2001 til April 13, 2002
Thu. April 4
Fri. April 5
Sat. April 6
Sun. April 7
Mon. April 8
Tue. April 9
Wed. April 10
Thu. April 11
Fri. April 12
Sat. April 13
Although we strive for accuracy in both copy and illustration, unintentional errors may occur.
We reserve all rights to correct errors.
We reserve the right to limit quantities.
Store Hours
Monday to Thursday 9 am - 8 pm
Friday 9 am - 8:30 pm
Saturday 8 am - 6 pm
Sunday 10 am - 4 pm
Steeles Ave. East
Gordon Baker
Mc Nicoll
Victoria Park
Nicey's
50 Morecambe Gate
Chester Le Blvd
Chester Le Blvd
Pharmacy
Finch Ave. East

NUFF NUFF NICE TINGS AT NICEY'S
This Week's Special
FRESH
Bone-In Chicken Breast
3.68 kg
1.67 lb
FRESH
Pork Loin Chops
5.45 kg
2.47 lb
GRADE AA AAA BEEF
Boneless Blade Roast or Steak
5.89 kg
2.67 lb
GRACE
White Rice
8 kg
3.97
Reg. $5.99
FRESH
Chicken Drumsticks
1.48 kg
.67 lb
GRADE AA AAA BEEF
Sirloin Tip Steak
6.99 kg
3.17 lb
Every Monday is Senior's Day
Get 10% Discount on Almost Every Item
Sorry! No Discount on Rice or Advertised Specials
Every Monday
MR. GOUDAS
Parboiled Rice
8 kg
4.97
Reg. $6.99
Supligen or Nutrament
Mix & Match
3/5.99
2.49ea
PLUS
Energy Drink
.99
LA MOLISANA, IMPORTED
Pasta
Assorted Cuts
450 g
.77
SEALTEST
2 % Milk
4 lt Bag
3.47
SEALTEST
100% Orange Juice
2 lt
1.67
SIFTO
Table Salt
1 kg
.77
CYNAR
Soft Drinks
Assorted 2 lt
.77
Please Turn Over For More Specials

SOLO PRACTICE-4

1. Use only $10.00 from your wallet to purchase the most quantity of items. Indicate the percent of the 10-dollar budget allocated to each category.
Responses will vary.

2. A senior citizen purchases 8 lbs. of Sirloin Tip Steak at 10% discount. Determine her change a 50-dollar bill.

Possible Solution
Cost before discount: 8 ×$3.17 = $25.36
Discount: 0.10 × 25.36 = $2.536
Cost to senior citizen: $25.36 – $2.536 = $22.82
Change: $50.00 –$22.82 = $27.18
Answer: The change is $27.18

PROBLEM-5
Black Friday

GROUP TASK-5

On Black Friday, Linda went to Flatbush Avenue in Brooklyn. There, she purchased on sale three sweaters: one black, one blue-grey, and one white-grey. The black sweater cost \$20.75. The blue-grey was \$1.25 less than the black. The white-grey sweater was as much as the black sweater and the blue-grey sweater combined. The tax is 8.75%.
Determine the change (if any) from a 100-dollar bill

Answers will vary.

Possible Solution:
Cost of the blue-grey sweater: \$20.75 –\$1.25 = \$19.50
Cost of the white-grey sweater: \$20.75 + \$19.50 = \$40.25
Cost of all sweaters before tax: \$20.75 + \$19.50 + \$40.25 = \$80.50
Tax on the sweaters: \$80.50 × 0.0875 = \$7.04
Total cost = \$80.50 + \$7.04 = \$87.54
Change out of 100-dollar bill: \$100.00 – \$87.54 =\$12.46
Answer: The change is \$12.46.

SOLO PRACTICE-5

1. Sue purchased one pair of pants for \$50.99, and one blouse that cost \$23.60 less than the pair of pants. She received \$21.62 as change. How much did Sue tender to the cashier?

Possible Solution
Cost of the blouse: \$50.99 –\$23.60 = \$27.39
Total cost (pants and blouse combined): \$50.99 + \$27.39 = \$78.38
Sue gave to the cashier: \$78.38 +\$21.62 = \$100.00
Answer: Sue tendered 100 dollars to the cashier to receive \$21.62 change.

2. Louis bought one shirt that cost as much as 3 pairs of socks combined. The price of one pair of socks was \$18.45. The tax was 8.75%. How much did he owe to the cashier?

Possible Solution
Cost of the shirt: 3× \$18.45 = \$55.35
(Note: Sue did not buy any socks)
Tax on the shirt: 0.875 × \$55.35 = \$4.84
Louis owed to the cashier: \$55.35 +\$4.84 = \$60.19
Answer: Louis owed to the cashier \$60.19.

PROBLEM-6
Taxi Ride

GROUP TASK-6

Last Sunday at 7:00 p.m., Denise hailed a taxicab for a 10-mile ride from Brooklyn to Manhattan. Inside the vehicle, the fare table below indicates the charges. In addition, Denise left a 15% tip for the driver.
Determine the amount she spent in all.

<u>Possible Solution</u>
Charge for one mile:
5 × = $3.75
Charge for the mileage:
10 × $3.75 =$37.50
Stoppage = $0.00
Weekday surcharge = $0.00
Weekend surcharge = $0.50
Total charge =
$3.50+$37.50+$0.50 = $41.50
Tip (15%) of total charge:
0.15 × $41.50 = $6.23
Denise spent in all:
$41.50 + $6.23 = $47.73
Answer: Denise spent in all $$47.73

Taxi Fare Table
Initial charge $3.50
Each 1/5 mile $0.75
Every 2-minute stop $0.40
Weekday surcharge (8 pm- 6 pm) $1.00
Weekend surcharge (4 pm- 8 pm) $1.00

SOLO PRACTICE-6

One weekday around 5:00 am, Jazzy took the same taxi from Queens to The Bronx for a 6-mile ride. At destination, she tipped the driver with a 5-dollar bill. Determine the amount Jazzy has spent in all on the 6-mile ride.

Possible Solution
Charge for one mile: 5 × = $3.75
Charge for the mileage: 6 × $3.75 =$22.50
Stoppage = $0.00
Weekday surcharge = $1.00
Weekend surcharge = $0.50
Total charge =$3.50 + $22.50 + $1.00 = $27.00
Tip = $5.00
Jazzy spent in all: $27.00 + $5.00 = $32.00
Answer: Jazzy spent in all 32 dollars.

PROBLEM-7
Bermuda Fish

GROUP TASK-7

A fisherman caught a giant 60-kg red snapper off the coast of Bermuda. He cut the fish into 4 sections. He sold Mid-Section B for $40, which was 5 times more than Mid-Section A. The Head-Section weighed 1/3 of the fish.
Complete the Bermuda Fish Table and the Bermuda Fish Circle Graph.
Write what you have observe in the table.

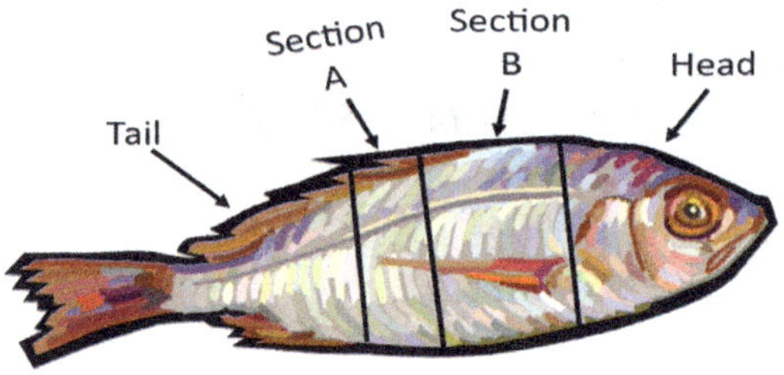

Bermuda Fish Table

Section	Weight	Fish Fraction	Notation in Decimal	Percentage	Cost
Head Section	20 kg	$\frac{20}{60} = \frac{1}{3}$	0.33	33%	$32
Mid-Sect. A	5 kg	$\frac{5}{60} = \frac{1}{12}$	0.08	8%	$8
Mid-Sect. B	25 kg	$\frac{25}{60} = \frac{5}{12}$	0.42	42%	$40
Tail Section	10 kg	$\frac{10}{60} = \frac{1}{6}$	0.17	17%	$16
Whole Fish	60 kg	$\frac{60}{60} = 1$	1.00	100%	$96

Observation
Answers will vary.

<u>Possible Response:</u> I (we) observe Mid-Section A and Mid-Section B of the fish combined are worth as much as the Head and the Tail combined. That is, each pair of sections is worth 50% of the weight and, inevitably. or proportionally, of the cost.

Bermuda Fish Circle Graph

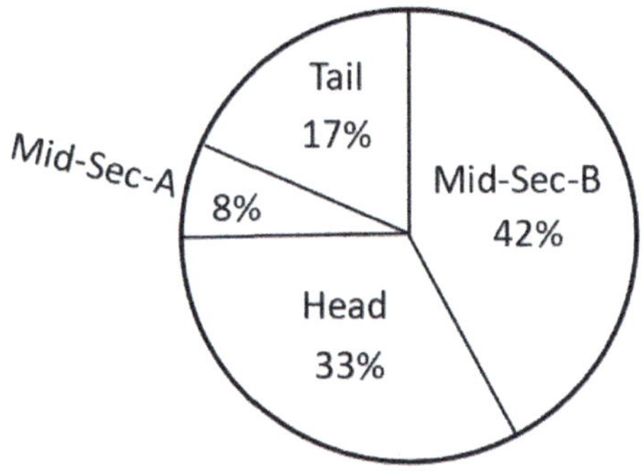

SOLO PRACTICE-7

If the total weight of the fish were 45 pounds, complete the table according to the proportions found in the Group Task Problem-5. The unit cost remains constant ($1.6 per lb.)

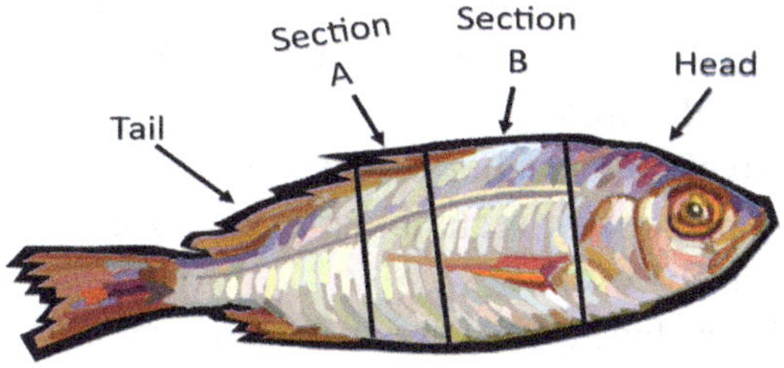

Bermuda Fish Table

Section	Weight	Fish Fraction	Notation in Decimal	Percentage	Cost
Head Section	14.85 lbs.	14.85/45	0.33	33%	$23.76
Mid-Section A	3.60 lbs.	3.60/45	0.08	8%	$5.76
Mid-Section B	18.90 lbs.	18.90/45	0.42	42%	$30.24
Tail Section	7.65 lbs.	7.65/45	0.17	17%	$12.24
Whole Fish	45 lbs.	45/45 = 1	1.00	100%	$72

Observation
Answer will vary

<u>*Possible Response:*</u>
(Similar to Group Task-5)

Bermuda Fish Circle Graph

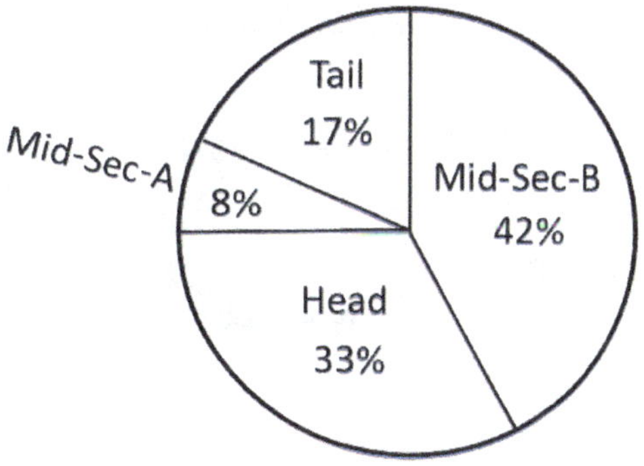

PROBLEM-8
Open Box

GROUP TASK-8

1. Given a regular sheet of paper or origami/oaktag, find the dimensions in cm (length and width). Then find and record the area.
2. Follow attentively the teacher's instructions to construct an open box or rectangular prism (see series of images below).
3. Determine the surface (total) area of the five faces of the open box, and round all areas to the nearest ten.
4. Compare this total area with the area of the original sheet of paper.
5. Complete the Open Box Table, the Front Face \$19.25 to be painted.
6. Complete the Open Boc Circle Graph.
7. What have you observed?

Folding the Sheet of Paper into an Open Box

A. Hotdog Fold

B. Hamburger Fold

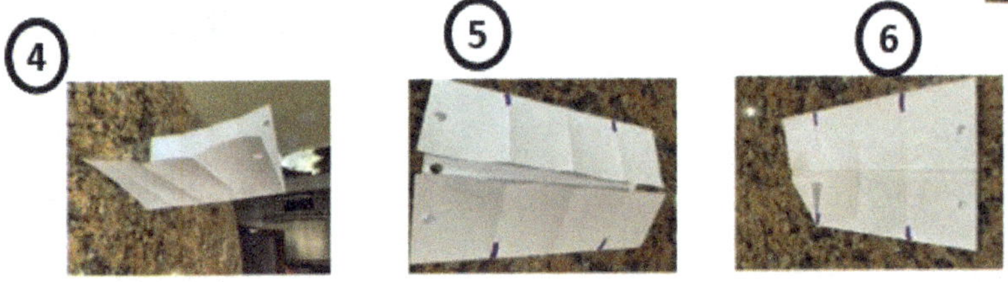

C. Prying opening Hamburger Fold

GROUP TASK-8
(continued)

Open Box Table					
Prism Face	Area	Fraction	Decimal	Percentage	Cost
Front	77 cm^2	$\frac{77}{603}$	0.1276	12.76%	$19.25
Back	77 cm^2	$\frac{77}{603}$	0.1276	12.76%	$19.25
Right	60.5 cm^2	60.5/603.3	0.1011	10.03%	$15.13
Left	60.5 cm^2	60.5/603.3	0.1011	10.03%	$15.13
Bottom	154 cm^2	$\frac{154}{603}$	0.2553	25.53%	$38.50
Hidden Area	174.3 cm^2	$\frac{174}{603}$	0.2889	28.89%	$43.58
Total Area (Orig. Paper)	603.3 cm^2	$\frac{603}{603} = 1$	1.0	100%	$150.84

Open Box Circle Graph

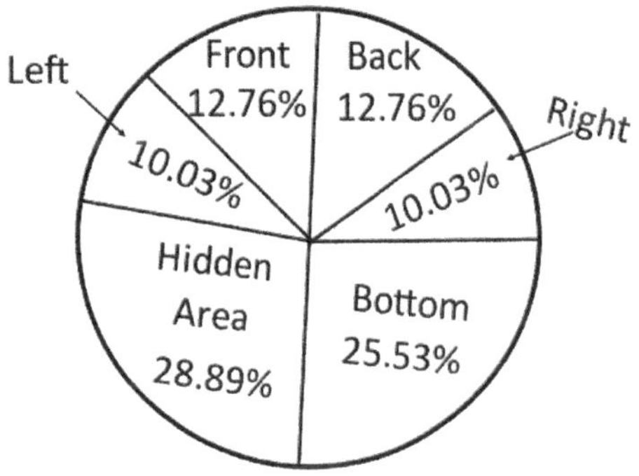

Observation
Answer will vary.

Possible response.
This activity has captivated us for many reasons. First, it was interesting to realized that we had started with an 8.5 in. × 11 in. sheet of paper. That is equivalent to 21.59 cm × 27.94 cm., yielding an area of 603.3 cm^2.
Amazingly, after all areas of the five faces have been computed and footed, the total is 429 cm^2, which is less than the area (603.3) of the original sheet of paper. So, this discrepancy is due to some area that has been folded into the inside of the box during the process of construction. This unseen area, the greatest of all, is referred to as Hidden Area in the table and circle graph. Representing approximately 30% of the area of the sheet of paper folded into a rectangular prism.

Moreover, it has been observed that the Bottom area (154 cm^2) is as twice as the Front area and the Back area combined. The Back and the Front are congruent and measure each almost 1/8 of the total area. Meanwhile, the Left Face and the Right Face are congruent for 10% each of the total area.
Note: It has been challenging to deal with the conversions of the fractions into conversions into decimals and percentages. So, some common-sense rounding of partial numbers has become necessary.

SOLO PRACTICE-8

Use an origami or oaktag sheet of paper of a size differing from the sheet used in Problem-8 Group Task. Fold the material similarly to the instructions given in the group activity. Construct a relevant table and circle graph.

Answers will vary.

PROBLEM-9
Prism-Cylinder Capacity

GROUP TASK-9

Revisit Prolem-8 (Open Box).
Reconstruct the box using origami paper or oaktag. Then construction a cylinder of your own dimensions that would have the same amount of cereal (capacity) as the prism (open box).
Justify your result (equal capacity) by pouring cereal in both containers.

Answer will vary.

<u>Possible response:</u>
After construction, the box or rectangular prism has the following dimensions:
length = 14 cm; width = 11 cm; and height = 5.5 cm.
The capacity or volume of the box is based on the formula:
$V = L \times W \times H$
$V = 14 \times 11 \times 5.5 = 857\ cm^3$

For the cylinder to contain the same amount of cereal, the capacity of the box must be set equaled to the capacity or volume of the cylinder. That is, the diameter of the cylinder must be computed based on our chosen of height of 8 cm by the formula for cylinder:

$V = \pi r^2 h$.
Let's replace V with 857 and h with 8.
$857 = 8\pi r^2$
$r^2 = 857 \div 8\pi$
$r^2 = 857 \div 25.12$
$r^2 = 34.11$
$r = \sqrt{34}$

$r = 5.83$ *cm.*

Therefore, the diameter is: 2 × 5.83. As to the construction, first with a compass opening to 5.83 cm, a circle ought to be traced representing the Base (area) of the cylinder. Using some oaktag or origami paper, the construction can be competed taken in consideration the height as 8 centimeters.
To justify the result, we must have filled the prism with rice and then pour the content into the cylinder.

SOLO PRACTICE-9

Construct a task similar to the previous Group Task of Problem 6B along with the a proposed solution.

Answers will vary.

PROBLEM-10
Library Proposal

GROUP TASK-10

The floor of a local library is to be renovated according to the sketch below. Examine carefully the data on the sketch. Complete the Table for Library Proposal. Complete the Circle Graph for Library Proposal. Write your observation.

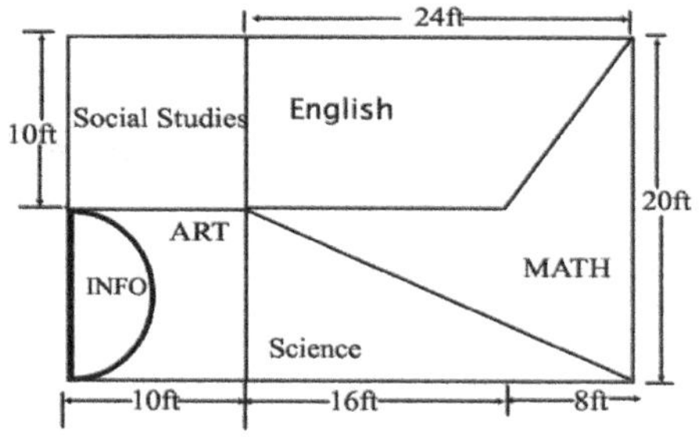

Table for Library Proposal					
Lib. Section	Area	Fraction	Decimal	Percentage	Cost
Social Studies	100 ft²	$\frac{100}{680}=\frac{5}{34}$	0.147	14.7%	$600
Science	120 ft²	$\frac{120}{680}=\frac{3}{17}$	0.176	17.6%	$720
English	200 ft²	$\frac{200}{680}=\frac{5}{17}$	0.294	29.4%	$1,200
Info	40 ft²	$\frac{40}{680}=\frac{1}{17}$	0.058	5.8%	$240
Art	60 ft²	$\frac{60}{680}=\frac{3}{34}$	0.088	8.8%	$360
Math	160 ft²	$\frac{60}{680}=\frac{4}{17}$	0.235	23.5%	$960
Total Area	680 ft²	$\frac{680}{680}=1$	1.0	100%	$4,080

Observation
Answers will vary.

Circle Graph for Library Proposal

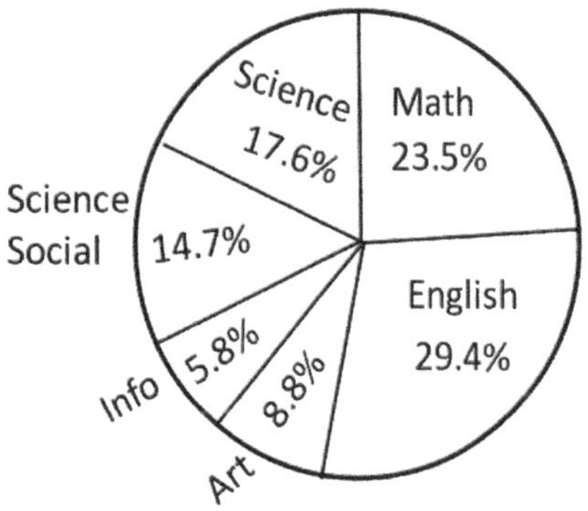

<u>Possible response</u>
We have observed that two rooms (English and Math combined) occupy more than 50% of the total space. Also, there are many ways to compute the area of each section. It was challenging to determine the area of the Art room because its shape is strange. But one the area (half-circle) of the Info is found, we subtract the area of the Info from the 10×10 square.

SOLO PRACTICE-10

Below is the blueprint of a City Shelter Building in case of emergency. Each room will be built at $500 per square yards.

- Calculate the total cost.
- Complete the City Shelter Building Table.
- Complete the City Shelter Circle Graph.

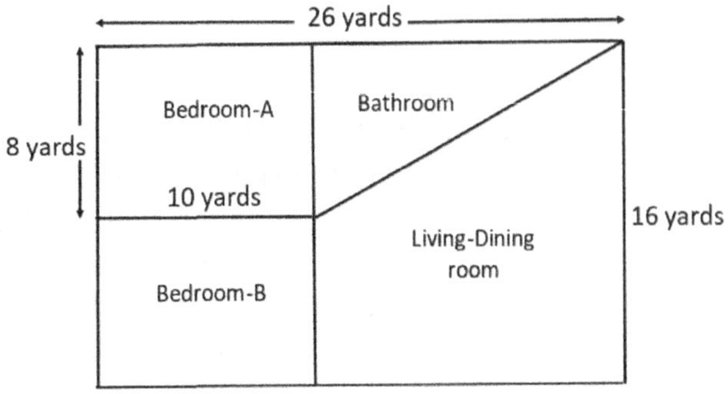

City Shelter Building Table					
Room	Area	Fraction	Decimal	Percentage	Cost
Bedroom A	80 yd²	$\frac{80}{416}=\frac{5}{16}$	0.1923	19.23%	$49,479.20
Bedroom B	80 yd²	$\frac{80}{416}=\frac{5}{16}$	0.1923	19.23%	$49,479.20
Bathroom	64 yd²	$\frac{64}{416}=\frac{2}{13}$	0.1538	15.38%	$39,583.36
Living-Dining Room	192 yd²	$\frac{192}{416}=\frac{6}{13}$	0.4615	46.15%	$118,750.08
Total Area	416 yd²	$\frac{416}{416}=1$	1.000	100%	$257,291.84

Observation
Answers will vary.

<u>*Possible response:*</u>
We have observed that the trapezoidal Living-Dining Room is the larger room. This is almost one-half or 46.15% of the shelter. The two rectangular bedroom areas are congruent, occupying each 19.23% of the total area. The triangular bathroom is the smallest at 15.38%.

City Shelter Building Circle Graph

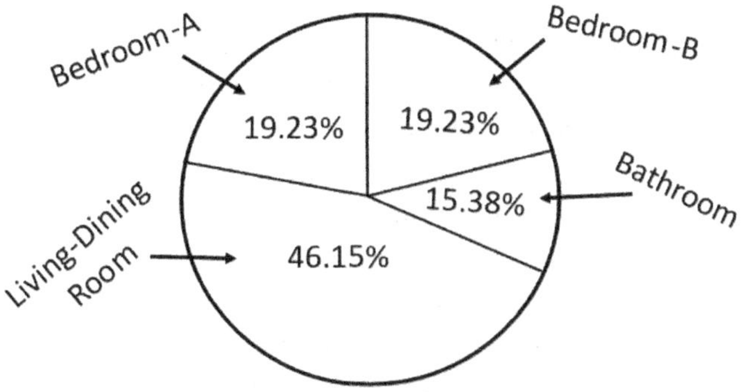

PROBLEM-11
Prime-&-Composite Numbers

GROUP TASK-11

Complete the 100-natural number chart below. Examine the chart and detect some patterns. Define prime numbers and color them red on the chart. Complete the Prime-and-Composite Numbers Table below. What have you observed?

1	2	3	4	5	6	7	8	9	10
11	12	13	14	15	16	17	18	19	20
21	22	23	24	25	26	27	28	29	30
31	32	33	34	35	36	37	38	39	40
41	42	43	44	45	46	47	48	49	50
51	52	53	54	55	56	57	58	59	60
61	62	63	64	65	66	67	68	69	70
71	72	73	74	75	76	77	78	79	80
81	82	83	84	85	86	87	88	89	90
91	92	93	94	95	96	97	98	99	100

Answers will vary.

<u>Possible response:</u>
A prime number admits only two factors. Extremely important, number 1 is neither a prime nor a composite because 1 has only one factor which is itself. But when a number has more than two factors, it is a composite. Bare in mind not to confuse prime numbers with odd numbers (not divisible by 2). Example: 2 is prime and pair; but 3 is prime and impair; 4 is composite and pair; 9 is composite and impair.

Prime-and-Composite Numbers Table

Natural Numbers	Quantity	Fraction	Decimal	Percentage
Prime	25	$\frac{25}{100} = \frac{1}{4}$	0.25	25%
Composite	74	$\frac{74}{100} = \frac{37}{50}$	0.74	74%
None	1	$\frac{1}{100}$	0.01	1%
Total	100	$\frac{100}{100} = 1$	1.0	100%

Prime-and-Composite Circle Graph

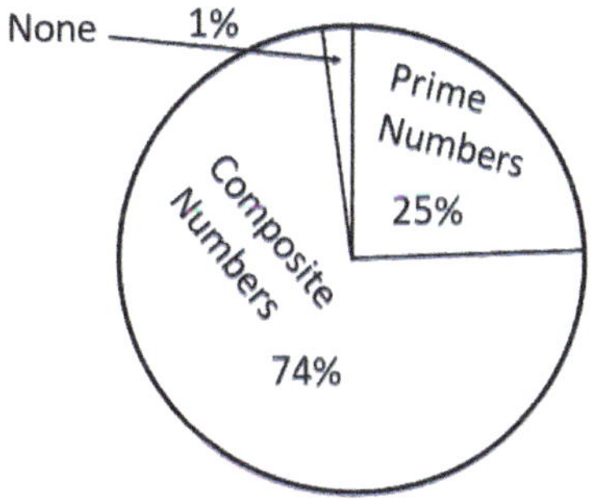

SOLO PRACTICE-11

Write the set N of all natural numbers, such that 100 < N <121.
Then circle all prime numbers.
Indicate the percentage of:
- Prime numbers
- Composite numbers
- None.

Possible solution:

(101) 102 (103) 104 105 106 (107) 108 (109) 110 111 112 (113) 114 115.

- *Prime numbers: 5/15 or 1/3 or 33.33%*
- *Composite numbers: 10/15 or 2/3 or 66.67%*
- *None: 0/15 or 0%.*

PROBLEM-12
Goose Play

GROUP TASK-12

One summer day, 630 brown geese and 370 white geese were found respectively on the West Bank and East Bank of the Huson River in Nyack. After a while, $\frac{2}{5}$ of the white geese swam to the west bank, and $\frac{3}{7}$ of the brown geese paddled to the east bank. Then 50% of the remaining white geese went westward, while 60% of remaining brown geese traveled eastward. Determine the number of each kind of geese on each bank of the Huson River by completing the accompanying table. Examine the table and write your observation.

West Bank			East Bank	
Geese White	Geese Brown	Hudson River	Geese White	Geese Brown
0	630		370	0
+148	-270		-148	+270
+111	-216		-111	+216
259	144		111	486

Observation
Answers will vary.

<u>Possible response:</u>
We observe that, at the beginning, the total number of geese (brown and white combined) is 1,000. And, after all the goose migrations from West Bank to East Bank, at the end, the total number of geese.

SOLO PRACTICE-12

Use a total number of 1,500 brown geese and white geese combined to build an solve a similar problem using the same proportions, fraction, and percentages among the varying number of geese.

The similar and new problem would be:

One summer day, 945 brown geese and 555 white geese were found respectively on the West Bank and East Bank of the Huson River in Nyack. After a while, $\frac{2}{5}$ *of the white geese swam to the west bank, and* $\frac{3}{7}$ *of the brown geese paddled to the east bank. Then 50% of the remaining white geese went westward, while 60% of remaining brown geese traveled eastward.*
Determine the number of each kind of geese on each bank of the Huson River by completing the accompanying table.
Examine the table and write your observation.

Possible Solution:

West Bank			East Bank	
Geese White	Geese Brown		Geese White	Geese Brown
0	945	Hudson River	555	0
+222	-405		-222	+405
+167	-324		-167	+324
389	216		166	729

Observation
Answers will vary

Possible response:
We observe that, at the beginning, the total number of geese (brown and white combined) is 1,500. And, after all the goose migrations from West Bank to East Bank, at the end, the total number of geese. Also, East Bank (895) has become more populous in geese than the West Bank (605 geese).

PROBLEM-13
Experimental-&-Theoretical Probability

GROUP TASK-13

Follow the teacher's instructions to conduct and compare an experimental probability and a theoretical probability, as follows:

1. First, conduct an experiment by tossing, at least 50 times, two dice of different colors and record the sums of numbers (outcomes) on a tally.
   Calculate the probability of each experimental outcome.
2. Second, complete the accompanying Addition Table and the related Sum-Probability Table. What have you observed?

Theoretical Probability (Addition Table)

Observation
Answers will vary.

Possible response:
We observe that there are varying degrees of chance or probability among the outcomes from the sum of the dice numbers. For example:

- P(Sum 1) = 0%
- P(Sum 2) = 2.8%
- P(Sum 3) = 5.6%
- P(Sum 4) = 8.3%
- P(Sum 5) = 11.1%
- P(Sum 6) = 13.5%
- P(Sum 7) = 16.7%
- P(Sum 8) = 13.8% .
- P(Sum 9) = 11.1%.
- P(Sum 10) = 8.3%
- P(Sum 11) = 2.8%.
- Sum 12 has 1% probability.

+	1	2	3	4	5	6
1	2	3	4	5	6	7
2	3	4	5	6	7	8
3	4	5	6	7	8	9
4	5	6	7	8	9	10
5	6	7	8	9	10	11
6	7	8	9	10	11	12

So, Sum 7 has the greatest probability with 16.7%.

Experimental Probability

Let's conduct the experimental probability to justify the theoretical probability. Differing outcomes may result from the various group experiment. However, the more the dice are cast, the more Sum 7 will have the greatest probability among the other sums.

SOLO PRACTICE-13

1. Determine the probability for one blue marble to be picked at random (without looking) from a sack that contains 4 white marbles and 6 yellow marbles.

Answers will vary.
<u>Possible response:</u>
Because there is no blue marble in the sack, the probability of picking up a blue marble is 0%.

2. A sack contains 4 white marbles and 6 yellow marbles.
   a). Find the probability for a yellow marble be picked at random on the first try.

Answers will vary.
<u>Possible response:</u>
Because there four 6 yellow marbles in 10, the probability of picking up one yellow marble on the first try is 6/10 or 60%.

   b). Determine the probability that a white marble be picked at random on the second try without replacement.

Answers will vary.
<u>Possible response:</u>
There were originally four 4 white marbles in 10, and that the yellow marble previously picked has not been replaced, leaving a total of 9. Therefore, the probability of picking up one white marble on the second try is 4 in 9, or 44%.

3. A sack contains 5 red marbles, 7 blue marbles, and 8 green marbles.
   a). Find the probability that a blue marble be picked at random on the first try.

Answers will vary.
<u>Possible response:</u>
Because there four 7 blue marbles in 20, the probability of picking up one blue marble on the first try is 7/20 or 35%.

   b). Determine the probability that a green marble be picked at random on the second try with replacement second try.

Answers will vary.
<u>Possible response:</u>
There were originally four 8 green marbles in 20, and that the blue marble previously picked has been replaced, leaving the same total of 20 for the second try. Therefore, the probability of picking up one green marble on the second try is 8 in 20, or 40%.

PROBLEM-14
Student-Work Analysis

GROUP TASK-14

Analyze the solution composed by the student (Paul Andre) to the problem below based on the accompanying rubrics. Indicate the score you would give Andre's work.

(Problem whose solution is to be analyzed):

A professional earns a gross pay of \$6,500. From that amount, $\frac{1}{5}$ is withheld for Federal taxes and 8% for local taxes. In addition, \$200 and \$150 are deducted respectively for future Medicare and Social Security benefits. After all deductions, the check indicates a net pay of \$ 4,330. Investigate whether the net pay is correct.

Student: Paul Andre | Class: 602 | Date: 02-28-22

Solution

Federal taxes: $\frac{1}{5} \times \$6{,}500 = \$1{,}300$

Local taxes: $\$6{,}500 \times 0.08 = \$510$

Total deductions:

$\$1{,}300 + \$510 + \$200 + \$150 = \$2{,}160$

Net pay: $\$6{,}500 - \$2{,}160 = \$4{,}340$

Answer: No. The pay check is not correct. It misses \$10.

Performance Levels	Rubric Performance Descriptors
Level 4	An appropriate strategy is used. For example, the student knows how to take 1/5 of \$6,500 and when to perform such basic operations as multiplication, addition, and subtraction. A solution line / phrase or label is provided for each operation. The process is coherent, and there is no mistake.
Level 3	An appropriate strategy is used. There may be one minor error at the beginning, with a consistent logic. The entire process is good, but the final answer may be wrong due to a computational error previously made. Some solution lines are missing.
Level 2	A good strategy was attempted. There are some errors of conceptual understanding leading to a wrong answer. Part of process may be good. Solution lines are missing.
Level 1	An inappropriate strategy is used. There multiple major errors some of which are of conceptual understanding and procedural fluency. Solution lines are missing or confusing.

Answers will vary.

After the students have read Andre's solution processes, no one knows for sure how their analyses of Andre's work would differ from each other. Consequently, what grade level they would attribute to this work remain to be seen.

However, it is hoped that student scorers would detect the follow mistakes:

Local taxes: \$6,500 × 0.08 = \$510.

But the right answer is \$520. Because the mistake was carried forward to the next solution line, it has inevitably impacted the subsequent wrong result of \$\$2.160. while the performance of the addition is correct. So, the net pay (\$4,340) found by Andre has derived from the originally wrong computation.

In conclusion, Andre has adopted an excellent strategy and followed it through. However, he has made some computational errors. For this reason, he deserves a Level 3.

SOLO PRACTICE-14

Compose a word problem and a related solution. Also, develop a set of rubrics that will enable the student-teacher to score the solution you have composed.

Answers will vary.

PROBLEM-15
Pattern Blocks

GROUP TASK-15

Explore your set of pattern blocks. Examine and identify the shape of each piece. Establish the relationships among them, and particularly the relationship (fraction, decimal and percentage) of each smaller polygon to the hexagon. Complete the table proportionally to the cost ($48) of the trapezoid. What have you observed?

Shape	Polygon	Fraction	Decimal	Percentage	Cost
	Trapezoid	$\frac{1}{2}$	0.33	33%	$43.05
	Hexagon	$\frac{6}{6} = 1$	1.0	100%	$86.10

Shape	Polygon	Fraction	Decimal	Percentage	Cost
	Triangle	$\frac{1}{6}$	0.17	17%	$14.35
	Hexagon	$\frac{6}{6} = 1$	1.0	100%	$86.10

Shape	Polygon	Fraction	Decimal	Percentage	Cost
	Square	$\frac{1}{3}$	0.33	33%	$28.70
	Hexagon	$\frac{6}{6} = 1$	1.0	100%	$86.10

Shape	Polygon	Fraction	Decimal	Percentage	Cost
	Rhombus	$\frac{1}{3}$	0.33	33%	$28.70
	Hexagon	$\frac{6}{6} = 1$	1.0	100%	$86.10

Observation
Answers will vary.

Possible response:
We observe that the rhombus and the square have an identical or same relationship with the hexagon. That is, each is 1/3 of the hexagon. Also, the square and the rhombus each is 2/3 of the trapezoid, which is one-half of the hexagon. Moreover, it takes six triangles to cover the entire hexagon the largest of all the polygons.

SOLO PRACTICE-15

Explore again your set of pattern blocks. Examine and identify the shape of each piece. Establish the relationships among them, and particularly the relationship (fraction, decimal and percentage) of each polygon to the rhombus, costing $60. Construct a series of tables similar to those on Problem-12 - Pattern Blocks-Group Task. What have you observed?

Shape	Polygon	Fraction	Decimal	Percentage	Cost
	Triangle	$\frac{1}{2}$	0.5	50%	$30
	Rhombus	$\frac{2}{2} = 1$	1.0	100%	$60

Shape	Polygon	Fraction	Decimal	Percentage	Cost
	Square	$\frac{2}{2} = 1$	1.0	100%	$60
	Rhombus	$\frac{2}{2} = 1$	1.0	100%	$60

Shape	Polygon	Fraction	Decimal	Percentage	Cost
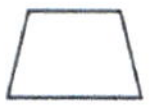	Trapezoid	$\frac{3}{2}$	1.5	150%	$90
	Rhombus	$\frac{2}{2} = 1$	1.0	100%	$60

Shape	Polygon	Fraction	Decimal	Percentage	Cost
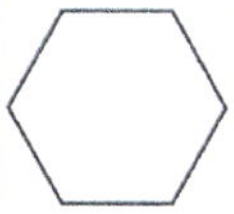	Hexagon	$\frac{6}{2} = 3$	3.0	300%	$180
	Rhombus	$\frac{2}{2} = 1$	1.0	100%	$60

Observation
Answers will vary.

Possible response:
It is mesmerizing to observe and express some of the relationships in terms of more than 100%. For example, the trapezoid is 150% or one time and on-half of the rhombus, and the hexagon is 300% or three times as much as the rhombus. That justifies the fact that some fractions can be greater than the unit considered, in this case the rhombus. Those are improper fractions.

PROBLEM-16
Coordinated Polygons

GROUP TASK-16

1. On the grid below, plot the points:
A(3,3); B(5,3); E(4,0); F(2,0)
Connect the points to get Polygon ABEF.
2. Plot D(6,0). Connect B, D, and E.
3. Plot C(6,6). Connect the points to get Polygon BCD.
4. Find the areas Polygons ABEF, BDE, BCD, and ABCDF.
5.Complete the table and the pie graph below, using the areas of the polygons.
6. Write your observation.

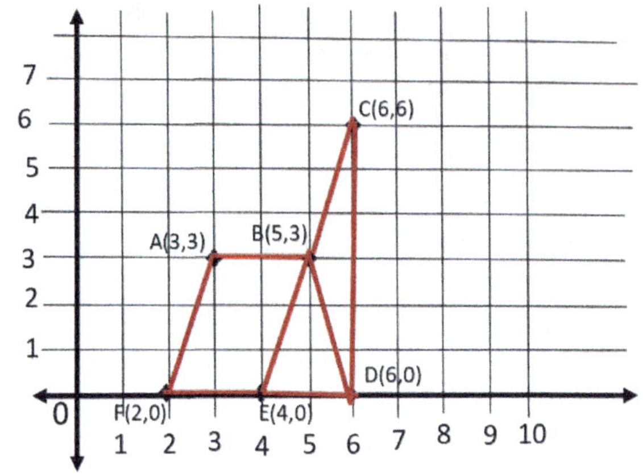

<u>Possible response:</u>
(After the plotting of the points and the construction of the polygons)
Area of Parallelogram ABEF = bh = 2 × 3 = 6 square units.
Area of Triangle BDE = 1/2bh = ½(2 × 3)= 3 square units.
Area of Triangle BCD = 1/2bh = ½(6 × 1) = 3 square units.
Area of Concave Pentagon ABCDF = $6\ u^2 + 3\ u^2 + 3\ u^2$ *= 12 square units.*

Coordinated Polygons Table					
Polygon	Area	Fraction	Decimal	Percentage	Cost
ABEF	$6\ u^2$	$\frac{6}{12} = \frac{1}{2}$	0.5	50%	$851
BDE	$3\ u^2$	$\frac{3}{12} = \frac{1}{4}$	0.25	25%	$425.50
BCD	$3\ u^2$	$\frac{3}{12} = \frac{1}{4}$	0.25	25%	$425.50
ABCDF	$12\ u^2$	$\frac{12}{12} = 1$	1.0	100%	$1,702

Coordinated Polygons Circle Graph

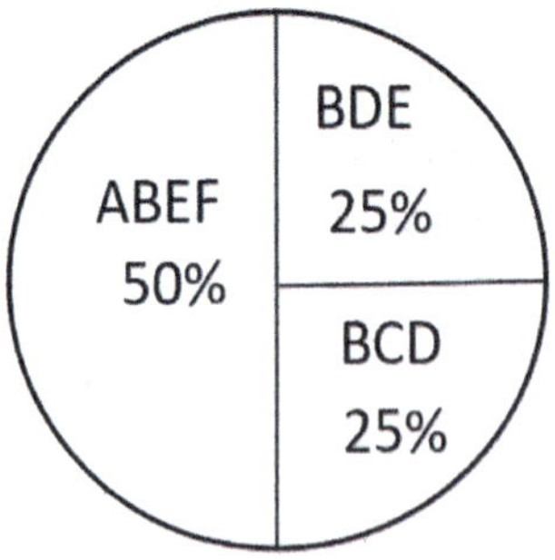

Observation
Answers will vary.

<u>Possible response:</u>
It is interesting to notice that the concepts of fraction, decimal, and percentage are also applicable to shapes as long as there is a partition of the whole. Particularly in this example, we have observed that the areas (6 u^2*) of the two triangles are congruent although their shape differ slightly, in terms dimensions. In fact, as to Triangle BDE the base is 2 units, and the height is 3 units. Whereas, in Triangle BCD the base is 6 units, and the height is 1 unit.*

SOLO PRACTICE-16

1. On the grid below, plot the points: A(3,3); B(5,3); F(4,0); G(2,0) Connect the points to get Polygon ABFG. Find the area.
2. Plot E(6,0). Connect B and E; then E and F. Find area of Polygon BEF.
3. Plot C(5,7) and D(6,7) Connect the points to get Polygon BCDE. Find the area.
4. Complete the Coordinated Polygon Table.
5. Write your observation.

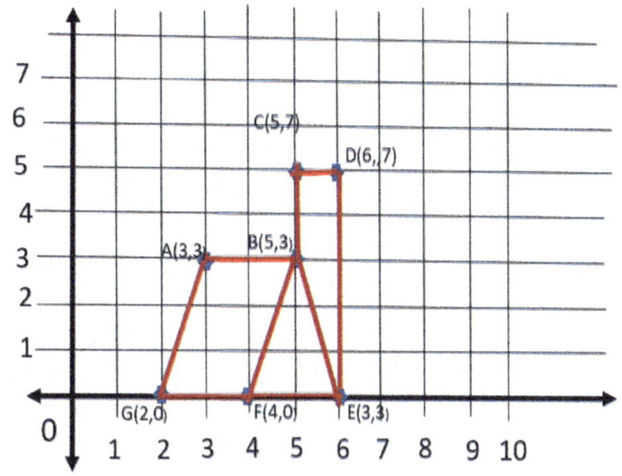

Observation
Answers will vary.

<u>*Possible response:*</u>
Instead of observation, I want to note the challenge in seeking to simplify the fractions whose numerators or denominators or involve a decimal point. For example, the fraction 6/14.5 can be transformed into 60/145 by multiplying both the numerator and the denominator by 10, and then simplified as 12/29 (or 0.41) by dividing both the numerator and the denominator by 5.

Similarly, the fraction 5.5/14.5 can be transformed into 55/145 by multiplying both the numerator and the denominator by 10, and then simplified as 11/29 (or 0.38) by dividing both the both the numerator and the denominator by 5.

Coordinated Polygons Table

Polygon	Area	Fraction	Decimal	Percentage	Cost
ABFG	6 u^2	6/14.5	0.41	41%	$162
BEF	3 u^2	3/14.5	0.21	21%	$81
BCDE	5.5 u^2	5.5/14.5	0.38	38%	$148.50
ABCDEG	14.5 u^2	14.5/14.5	1.0	100%	$391.50

Coordinated Polygons Circle Graph

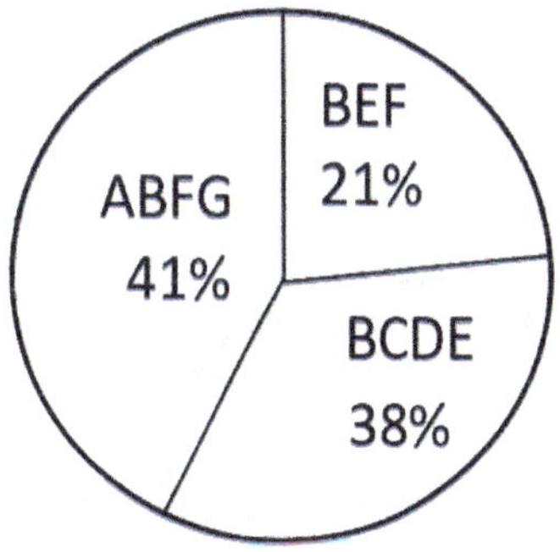

PROBLEM-17
What's π?

GROUP TASK-17

Conduct the following experiment including your findings as to the meaning of π..

1. Use some household objects to trace four circles A, B, C, and D of different sizes.
2. Trace the diameter of each circle.
3. Select dried beans of the same size and glue them on the circumference of the circle. Record that number of beans.
4. Glue beans similarly on the diameter of each circle, and record that number.
5. Use these new-found data to compete the What's Pi Table.
6. Analyze the table and report your findings.

<u>*Possible response:*</u>
(After the beans have been placed on the circumference and diameter of each circle, the experiment table will look like this:

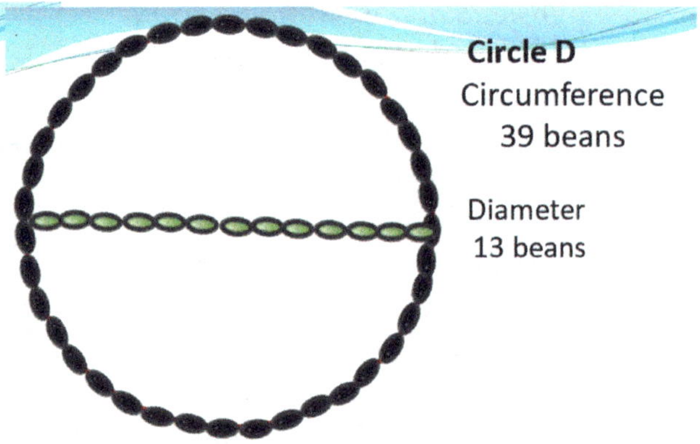

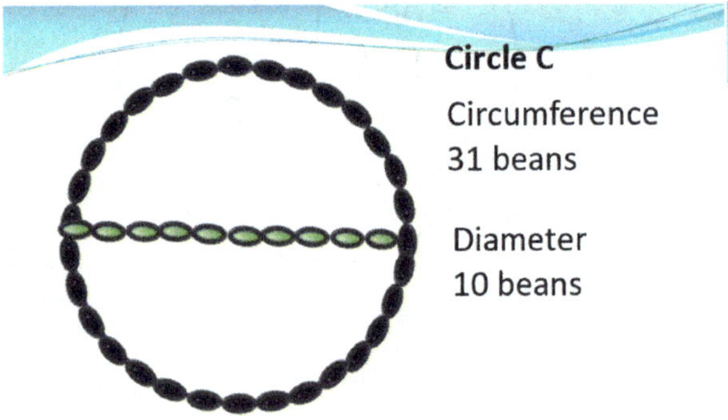

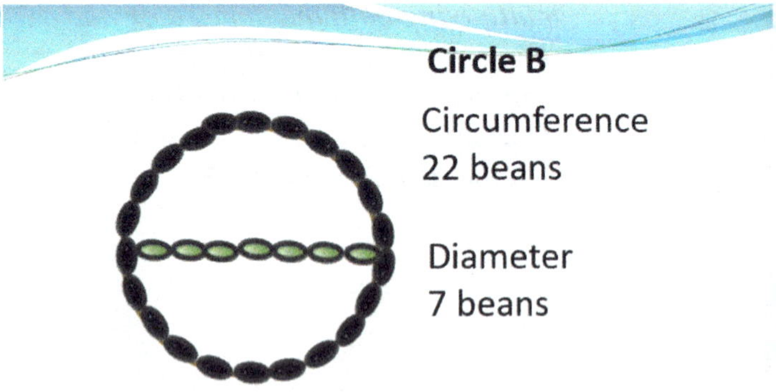

GROUP TASK-17
(continued)

What's Pi Table?			
Circles	Circumference	Diameter	Circumference to Diameter Ratio
Circle A	15.5	5	15.5/5 = 3.1
Circle B	22	7	$\frac{22}{7}$ = 3.1428
Circle C	31	10	$\frac{31}{10} = 3.1$
Circle D	39	13	$\frac{39}{13} = 3$

The findings will vary.
One group of students may find that the ratios of the circumferences of the circles to their respective diameters differ. Several reasons may explain these discrepancies. First, the beans used might be of varied lengths. Second, some gaps exist between beans. Third, some circles are not round enough.
For these reasons, it becomes necessary to determine the mean or average of the four ratios above.
Average ratio = (3.1 + 3.14 + 3.1 + 3) ÷ 4 = 3.08

In conclusion, the average ration can be taken as the ratio of the experiment. It indicates that the on average the ratio of circumferences of the circles to their circumferences is 3.08. That is the circumference of any circle is three times as long as the diameter of the same circle. That constant ratio is the meaning of Pi (π).

Ancient Hreek mathematician Archimedes of Syracuse was credited for the discovery of Pi. With a perfectly round circle, ratio 22/7 is the best representation of π. Because that division never ends and the next digit in the quotient is unpredictable, π is known as rational number. The symbol π for was first used in 1706 by British mathematician William Jones, and popularized by Swiss mathematician Leonhard Euler.
π = 3.141592653589793238...

SOLO PRACTICE-17

Repeat a similar experiment with as many circles as you wish to prove the value of π.

Answers will vary.

PROBLEM-18
Magic Tangram

GROUP TASK-18

Using the seven pieces of your Tangram set to complete the task below.

1. Explore your Tangram set.
2. Tessellate the pieces into a larger square. Explain how you have arrived at doing it.
3. Determine the area of each polygon to the nearest mm^2, using many strategies.
4. Complete the Magic Tangram Table and the Magic Tangram Circle Graph.
5. Write your observation.

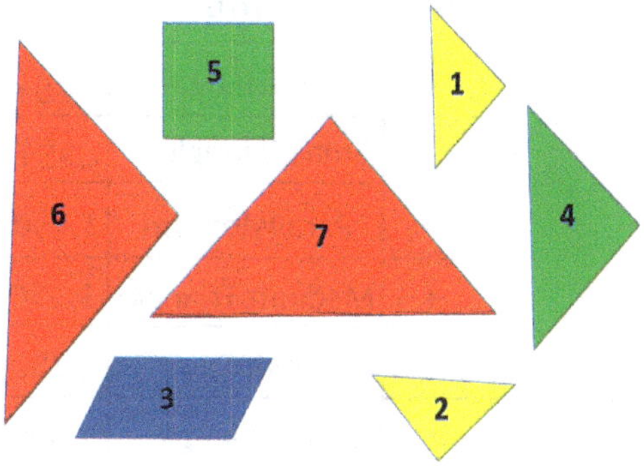

Answers will vary.
Considering the exploration, we have realized that there are some striking relationships among the polygons. For example, two smallest triangles cover the area of the parallelogram, the area of the medium triangle and the area of the square. But it takes 8 smallest triangles to cover the area of one of the largest triangles.

<u>Possible solutions:</u>
While exploring and measuring all the polygon, we found that the base and the height are equal for the medium right triangle:
Dimensions of medium triangle: base = 5 cm or 50 mm; height = 5 cm or 50mm
Area of medium triangle = $1/2bh = \frac{1}{2}(50 \times 50) = 1{,}250\ mm^2$

Based on the area of the medium triangle, the areas of the other polygons can be computed using proportions or the relationships among their sizes or areas described earlier.
Area of the square $=1{,}250\ mm^2$
Area of the parallelogram $=1{,}250\ mm^2$
Area of one of the smallest triangles $= 1{,}250\ mm^2 \div 2 = 625\ mm^2$
Area of one of the largest triangles $= 1{,}250\ mm^2 \times 2 = 2{,}500\ mm^2$

With this data we can complete the table and circle graph below.

PROBLEM-18
Magic Tangram
(continued)

Magic Tangram Table						
Polygon	Name	Area	Fraction	Decimal	Percentage	Cost
1	Small Triangle	625 mm^2	$\frac{625}{10000} = \frac{1}{16}$	0.0625	6.25%	$218.75
2	Small Triangle	625 mm^2	$\frac{625}{10000} = \frac{1}{16}$	0.0625	6.25%	$218.75
3	Parallelogram	1,250 mm^2	$\frac{1250}{10000} = \frac{1}{8}$	0.125	12.5%	$437.50
4	Medium Triangle	1,250 mm^2	$\frac{1250}{10000} = \frac{1}{8}$	0.125	12.5%	$437.50
5	Square	1,250 mm^2	$\frac{1250}{10000} = \frac{1}{8}$	0.125	12.5%	$437.50
6	Large Triangle	2,500 mm^2	$\frac{2500}{10000} = \frac{1}{4}$	0.25	25%	$875.00
7	Large Triangle	2,500 mm^2	$\frac{2500}{10000} = \frac{1}{4}$	0.25	25%	$875.00
Total	xxxxxxxxxx	10,000 mm^2	$\frac{10000}{10000} = 1$	1.0	100%	$3,500.00

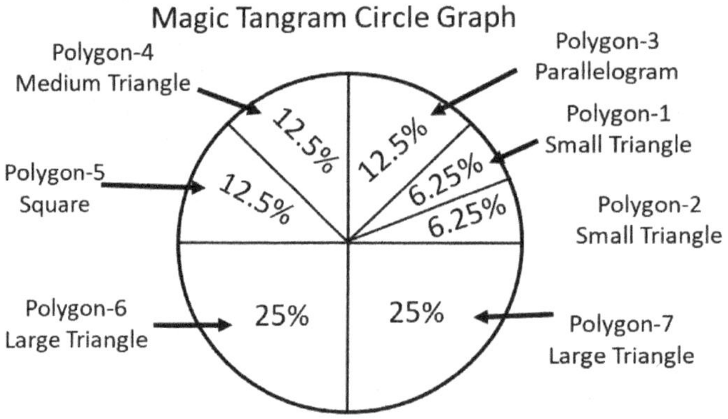

Observation

<u>*Possible response:*</u>
We have observed that the relationships or ratios among the areas of the polygons remain constant in all the columns of the table, as detected during the exploration phase of the activity. That is, when all the Tangram pieces have been tessellated into a larger square representing the whole, the two largest congruent triangles make up 50% of the structure.
Moreover, areas of the parallelogram, the square and the medium triangle are congruent, representing 12.5% or 1/8 of the structure. Finally, the two smallest congruent triangles combined are 6.25% of the entire structure or ½ of the parallelogram.

SOLO PRACTICE-18

From the original 7 pieces of the Tangram Select any set of five pieces of the Tangram set, remove the two largest triangles. Tesselate the rest of the polygons into a jumbo triangle. and construct a table and circle graph similar to the ones in the previous Group Task. Use $40 as the cost of the square.
What have you observed in comparison to the previous table and circle graph constructed during the previous Group Task.

Magic Tangram Table

Polygon	Name	Area	Fraction	Decimal	Percentage	Cost
1	Small Triangle	625 mm^2	$\frac{625}{5000}=\frac{1}{8}$	0.125	12.5%	$20
2	Small Triangle	625 mm^2	$\frac{625}{5000}=\frac{1}{8}$	0.125	12.5%	$20
3	Parallelogram	1,250 mm^2	$\frac{1250}{5000}=\frac{1}{4}$	0.25	25%	$40
4	Medium Triangle	1,250 mm^2	$\frac{1250}{5000}=\frac{1}{4}$	0.25	25%	$40
5	Square	1,250 mm^2	$\frac{1250}{5000}=\frac{1}{4}$	0.25	25%	$40
Total	xxxxxxxxxx	5,000 mm^2	$\frac{5000}{5000}=1$	1.0	100%	$160

Magic Tangram Circle Graph

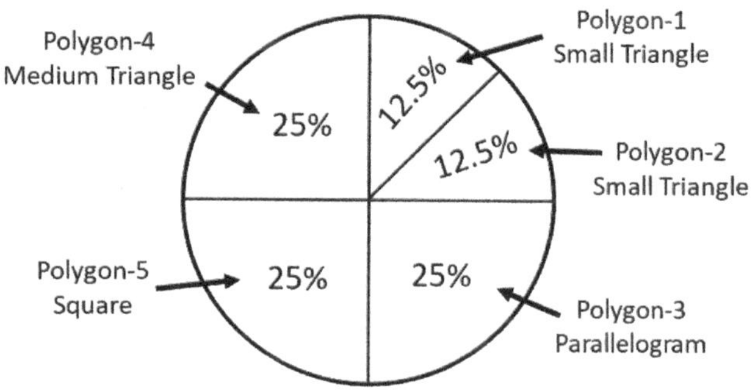

Answer will vary.

<u>*Possible response.*</u>
There is no need to recalculate the area of the polygons. Once the two largest triangles have been removed carefully, the five other pieces remain tessellated for a total area of 5,000 mm^2. So, the ratio of each polygon in the new structure which is a jumbo triangle has conversely doubled as indicated in the table and circle graph above. It proves the fact that dividing the denominator of a fraction leads to doubling the value of this fraction. That why, for example the square that was worth 12.5% of 10,000 mm^2 is now worth 25% of 5,000 mm^2. This observation is similar for each of the five pieces of the new structure.

PROBLEM-19
Best Shooter

GROUP TASK-19

One summer, five youngsters Rich, Jazz, Rick, Shy, and Fah formed a household basketball team in Nyack. Before confronting a neighborhood team, they decided to find out who among them was the best shooter. To do so, they agreed to a shootout competition, yielding the following results in shot made and shots attempted.

1. Determine the best shooter while completing the Best Shooter Table.
2. Rank all the shooters.
3. Find the team average.

Best Shooter Table

Shooters	Shots Made	Shots Attempted	Fraction	Decimal	Percentage
Rich	9	12	$\frac{9}{12}=\frac{3}{4}$	0.75	75%
Jaz	11	15	$\frac{11}{15}$	0.73	73%
Rick	14	20	$\frac{14}{20}=\frac{7}{10}$	0.7	70%
Shy	8	10	$\frac{8}{10}=\frac{4}{5}$	0.8	80%
Fah	11	14	$\frac{11}{14}$	0.78	78%

Ranking: 1. Shy 80% 2. Fah 78% 3. Rich 75% 4. Jazz 73% 5. Rick 70%

Average team performance: $(0.75 + 0.73 + 0.7 + 0.8 + 0.78) \div 5 = 0.752$ or 75.2%

Observation
Answers will vary.

<u>*Possible response:*</u>
It is interesting to observe that although Shy has the least number of shots made, she ended up being the best shooter. For the made shooter is not determined by the number of shoots that went through the net, but by the greatness of the ratio of shot made to shots attempted. Look at Rick! He had 14, the greatest number of shots. However, he was last in the ranking because the number of his shots attempts has drowned his performance overall. The same analysis can be made for all the other shooters' scores.

SOLO PRACTICE-19

Examine the score table below. Based on your analysis, determine who is the most prolific shooter.

Sample Response

Best Shooter Table					
Shooters	Shots Made	Shots Attempted	Fraction	Decimal	Percentage
Ann	12	16	$\frac{12}{16} = \frac{3}{4}$	0.75	75%
Max	10	14	$\frac{10}{14} = \frac{5}{7}$	0.714	71.4%
Sue	10	12	$\frac{10}{12} = \frac{5}{6}$	0.833	83.3%
Tom	9	16	$\frac{9}{16}$	0.563	56.3%
Fey	11	13	$\frac{11}{13}$	0.846	84.6%

Answers may vary.

Possible response:
Based on the analysis of the table, Fey, at 84.6%, is the best shooter, followed by Sue (83.3%), Ann (75%), Max (71.4%), and Tom (56.3%).

PROBLEM-20
Mango Juice

GROUP TASK-20

Narrator: Ladies and gentlemen, please welcome Jeanne and Mary. Jeanne is a young lady who loves mango juice. Mary is a peddler who wants to sell as many mangoes as possible. This morning out goes Jeanne in search of her favorite fruit. As Jeanne crosses a street corner, a soft voice breaks the silence in her mind.

Mary: My dear beautiful lady, please come and check my mangoes.
Jeanne: Okay. No problem. Indeed, I need some mangoes to make juice.
Mary: That's a good idea. Today they're fresh and delicious.
Jeanne: How do you sell these Francique mangoes?
Mary: Three for five dollars.
Jeanne: Okay. No problem. I'll take eight.
Narrator: Freeze! How much does Jeanne owe for the eight mangoes?
Mary: Mmm! I'll see….

$$(\$5 \div 3) \times 8 = \$18.33$$

Jeanne: I also need some Cinnamon mangoes.
Mary: That's a good idea. Today they're sweet and juicy.
Jeanne: Is it the same price?
Mary: Nope! These are five for three dollars.
Jeanne: Okay. No problem. I'll take twelve.
Narrator: Freeze! How much does Jeanne owe for the 12 mangoes?
Mary: Mmm! I'll see….

$$(\$3 \div 5) \times 12 = \$7.20$$

Jeanne: That's all…Here's a crispy twenty-dollar bill…Keep the change.
Narrator: Freeze! How much change has Jeanne left for Mary? Explain.

Possible explanation:
Jeanne has left nothing for Mary. Instead, the money is not enough. The total amount owed is \$20.53. That is, \$18.33 for the Francique mangoes plus \$7.20 for the Cinnamon mangoes. On the contrary, Jeanne is missing \$0.53.

SOLO PRACTICE-20

Re-read the playlet *Mango Juice*, then answer the following questions.

Possible response:

1. Did Jeanne and Mary know each other prior to the encounter?
   Provide some textual evidence.
   No Jeanne and Mary did not know each other prior to the encounter. In the Playlet, Mary addresses to Jeanne as "My dear beautiful lady."

2. How does Mary advertise her mangoes?
   Mary uses a series of adjectives to characterize her mangoes. In fact, she boasts that her mangoes are fresh. Delicious, sweet, and juicy.

3. Which mango brand is more expensive (per unit price)? How do you know?
   The Francique mango brand is more expensive because the unit price is $1.67, whereas the Cinnamon mango brand is $0.60. The unit price is found by dividing the cost by the quantity of mangoes for that cost.

4. How would you describe the temperament of each character in the Playlet?
   Both Mary and Jeanne appear very courteous and respectful. Also, they seemingly want to negotiate in good faith. However, Jeanne may have mistakenly believed she had enough money, when she did not. No one knows for sure whether she was trying to be deliberately deceitful or not.

5. Indicate the idiosyncratic response of the vendor.
   The vendor has the habit of saying: Mmm! I'll see…I's a good idea…"

6. Indicate the idiosyncratic intervention of the narrator.
   The narrator introduces her intervention with the word: "Freeze" before posing a math question pertinent to the particular aspect of the transaction between Mary and Jeanne.

7. Does Jeanne have enough money for the purchase? How do you know?
   No. Jeanne did not have enough money because all she had was $20. Whereas the total cost of both mango brands amounted to $23.53.

8. Make a list of all the adjective in the text and indicate their meaning.
   Adjective from the text: dear, beautiful, some, good, fresh, delicious, these, three, five, eight, sweet, juicy, twelve, crispy.

9. How would you characterize Jeanne?
   Jeanne can be characterized as a lover of mango juice. Also, she appears polite and agreeable. But she does not have a good grasp about money, if not purposefully tricky.

10. Compose your own playlet to share with the class. *Answer will vary.*

PROBLEM-21
Phone Sale

GROUP TASK-21

In the 1990's, two telephone companies Flashy and Speedy were competing in New York. Flashy would charge 50 cents per hour in addition to a 3-dollar flat fee for service. Whereas, Speedy would charge one dollar per hour and no service fee.

1. For what quantity of hours would the total charge be the same, regardless of the company?
2. Which store would you patron? Why?
3. Write an equation for each potential transaction at Flashy and Speedy.

Answers will vary.
Possible solutions.
Multiple strategies can be used to find out the quantity of hours for which the total charge would be the same at either store.
For example, a table like this one below can be useful.

Phone Sale Table		
Hours	Total Charge Flashy	Total Charge Speedy
0	($0.50 × 0) + $3 = $3.00	($1.00 × 0) = $0.00
1	($0.50 × 1) + $3 = $3.50	($1.00 × 1) = $1.00
2	($0.50 × 2) + $3 = $4.00	($1.00 × 2) = $2.00
3	($0.50 × 3) + $3 = $4.50	($1.00 × 3) = $3.00
4	($0.50 × 4) + $3 = $5.00	($1.00 × 4) = $2.00
5	($0.50 × 5) + $3 = $5.50	($1.00 × 5) = $5.00
6	($0.50 × 6) + $3 = $6.00	($1.00 × 6) = $6.00
7	($0.50 × 7) + $3 = $6.50	($1.00 × 7) = $7.00

Another strategy, Algebra, may be chosen by upper-grade (7th, 8th and 9th) students engaging in generating a system of equations representing the two potential transactions from Flashy and Speedy, and then solve the unknown (variable) x.

For example: $y = 0.5x + 3$*, representing the monthly charge at Flashy;*
$y = x$*, representing the monthly charge at Speedy*
Let's solve for:

$0.5x + 3 = x$
$3 = x + 0.5\,x$
$6 = x.$

__Answer.__ For 6 hours, it does not matter where you

SOLO PRACTICE-21

A. Write in algebraic form.

1. Number n increased by 7.
   $n + 7$
2. Number 8 decreased by z.
   $8 - z$
3. The sum of the variables a and b is twenty-four.
   $a + b = 24$
4. The difference between m and p is 0.
   $m - p = 0$
5. A quotient of t and q, decreased by the sum of r and s.
   $(t \div q) - (r + s)$
6. The square of c equals the sum of the square of a and the square of **b**.
   $c^2 = a^2 + b^2$
7. The product of the sum of a and b, and the difference between a and b is fourteen.
   $(a + b)(a - b) = 14$

B. Solve the following equations.

1). $10x = 56$

$x = 56/10$

$x = 5.6$

2). $3p - 12 = -18p + 93$

$3p + 18p = 03 + 12$

$21p = 105$

$p = 5$

3). $a - b = 4$, and $a + b = 36$

$a = (b + 4) \quad a = 36 - b$

$b + 4 = 36 - b$

$b + b = 36 - 4$

$2b = 32$

$b = 32 \div 2$

$b = 16$

$a = b + 4$

$a = 16 + 4$

$a = 20$

4). $\frac{1}{2}m + \frac{1}{3}m + 17 = 20 - \frac{1}{6}m$

$\frac{1}{2}m + \frac{1}{3}m + \frac{1}{6}m = 20 - 17$

$\frac{3}{6}m + \frac{2}{6}m + \frac{1}{6}m = 3$

$\frac{6}{6}m = 3$

$m = 3$

PROBLEM-22
Metro Card

GROUP TASK-22

In February 2016, in New York City one single-ride MTA MetroCard cost \$2.75. A Middle School student swiped the card only to go to school or a part-time job and return home. At the same time, the one-month MetroCard is available for \$90.

1. Determine the discount, in dollars, a rider would benefit if the young strapnager purchased a MTA one-month fare in February 2016.
2. Determine the percent discount.

Note: The student used his card throughout the month, excepts on Saturdays and Sundays.

Answers will vary.

Possible response.

The month of February 2016 fell in a leap year, meaning that month carries 29 days. According to the calendar, the last day is a Monday. So, the straphanger must have utilized the MTA card for 21 days when removing Saturdays and Sundays.

1. Monthly cost when using single rides:

$2 \times \$2.75 \times 21 = \$115.50$

2. Discount in dollars if using the one-month fare:

$\$115.50 - \$90.00 = \$25.50$

Discount in terms of percentage:

$[(115.50 - \$90.00) \div 115.50] \times 100 \approx 22.08\%$

SOLO PRACTICE-22

Referring to to Problem-19, the Metropolitan Transit Autority plans to rise the fare in the following month (March 2016) by 6% for the single rides, while keeping the one-month fare constant. What would be the percent discount in March 2016 if the same rider utilized the one-month MetroCard.
Note: The rider would use his card twice a day and would stay home on Saturdays and Sundays.

Answers will vary.

<u>Possible response:</u>
March 2016 carries 27 weekdays, plus 8 Saturdays and Sundays combined.
With a 6% increase, the single ride in March 2016 would be:
$\$2.75 \times 1.06 = \$2.92$
Monthly cost when using single rides:
$2 \times \$2.92 \times 27 = \$157.68$
Discount in terms of percentage:
$[(157.68 - \$90.00) \div 157.68] \times 100 \approx 42.92\%$

PROBLEM-23
Water Flow

GROUP TASK-23

A cylindrical water container is one-half meter tall with a diameter of 1 foot. When the faucet is activated, the flow flows at a rate of 5.209 liters per minute. At this rate of water flow, determine the number of minutes it would take for the container to be emptied.
Note: 1 dm^3 = 1 liter
1 foot = 30.48 centimeters
$\pi = 3.14$

Answers will vary.

Possible response:
Height = ½ m = 0.5 m = 5 dm
Diameter = 1 foot = 30.48 cm = 3.048 dm
Radius = (3.048) ÷ 2 = 1.524 dm.
Capacity of the container:
$V = \pi r^2 h$
$V = 3.14 \times (1.524)^2 \times 5$
$V = 36.4644\ dm^3$
Capacity = 36.464 liters
Time to get the container emptied:
$36.464 \div 5.209 = 7.00$ *minutes*
Answer: At this rate, it would take the container 7 minutes to be emptied.

SOLO PRACTICE-23

A huge cylindrical container is ¾ meter tall with a diameter of 2 foot. When the faucet is activated, it drops 2.188 liters per minute. At this rate, determine the number of hours (and minutes if any) it would take for the container to be completely emptied.

Note: $1\ dm^3 = 1$ liter

1 foot = 30.48 centimeters

$\pi = 3.1416$

Answers will vary.

Possible response:

Height = ¾ m = 0.75 m = 7.50 dm

Diameter = 2 feet = 2 × 30.48 cm = 6.096 dm

Radius = (6.096 dm) ÷ 2 = 3.048 dm.

Capacity of the container:

$V = \pi r^2 h$

$V = 3.1416 \times (3.048)^2 \times 7.50$

$V = 218.898\ dm^3$

Capacity = 218.898 liters

Time to get the container emptied:

218.898 ÷ 2.188 = 100 minutes

100 minutes = 1 hour 40 minutes.

Answer: At this rate, it would take the container 1 hour 10 minutes to be emptied.

PROBLEM-24
SWEET LEMONADE

GROUP TASK -24

This morning, Mom made some lemonade that was found delicious by all children: Julienne, Shandra, Jude, Angelo, and Myriam. Mom's recipe is the following: 2 cups of water, 6 spoons of sugar, 2.5 lemons, and ¾ teaspoons of vanilla extract. Complete the Sweet Lemonade Table. What have you returned?

Answers will vary.

<u>*Possible solutions:*</u>
One way to find the ingredients that are unknown is to use the Law of Proportion. That is, in all proportions the product of the means equals the product of the extremes. a/b = c/d.

Sweet Lemonade Table				
People	Water	Sugar	Lemons	Vanilla
Mom	4 cups	6 spoons	2.5 lemons	$\frac{3}{4}$ T.spoon
Julienne	8 cups	12 spoons	5 lemons	$1\frac{1}{2}$ T. spoons
Shandra	2 cups	3 spoons	1.25 lemons	$\frac{3}{8}$ T. spoon
Jude	16 cups	24 spoons	10 lemons	$2\frac{1}{4}$ T.spoons
Angelo	20 lemons	30 spoons	12.5 lemons	$3\frac{3}{4}$ T.spoons
Myriam	40 spoons	60 spoons	25 lemons	$7\frac{1}{2}$ T. spoons

SOLO PRACTICE-24

Based on what you have learned from the previous activity, complete the table below.

Cake Recipe Table				
People	Flour	Sugar	Milk	Vanilla
Nordie	5 lbs.	1 cup	2.5 cans	$1\frac{1}{2}$ spoons
Guerda	10 lbs.	2 cups	5 cans	3 spoons
Claude	22 lbs.	4.4 cups	11 cans	$6\frac{3}{5}$ spoons
Marly	45 lbs.	9 cups	22.5 cans	$13\frac{1}{2}$ spoons
Farah	56 lbs.	11.2 cups	28 cans	$16\frac{4}{5}$ spoons
Rose	2.5 lbs.	0.5 cup	1.25 cans	$\frac{3}{4}$ spoon

PROBLEM-25
Diamond Park

GROUP TASK -25

A private park (see sketch below) is to be renovated. The diagonals of the smaller rhombus measure 60 yards and 80 yards, whereas the diagonals of the larger rhombus are 180 yards and 240 yards.
Red roses are to be planted inside the smaller rhombus at \$100.25 per yd^2, whereas green grass will grow in the rest of the park at \$10.00 yd^2. Finally, the park will be secured will a fence at \$5 per ft^2.
Determine the total cost of the park renovation.
Complete the table and the pie chart.
What have you observed?

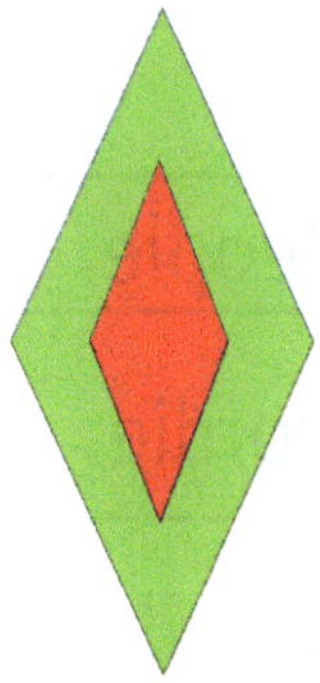

Diamond Park Budget Pie Chart

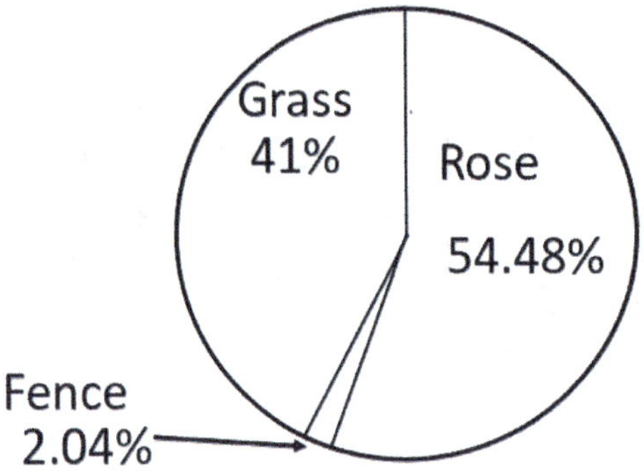

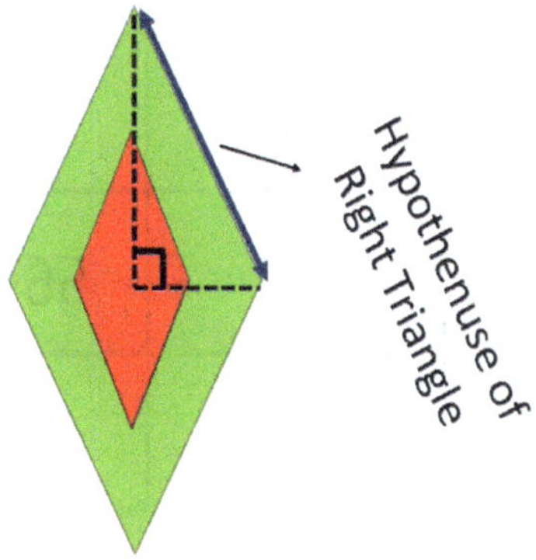

Diamond Park Budget Table				
Park Section	Cost	Fraction	Decimal	Percentage
Rose	\$240,600	$\frac{240,600}{441,600} = \frac{401}{736}$	0.5448	54.48%
Grass	\$192,000	$\frac{192,000}{441,600} = \frac{10}{23}$	0.4348	43.48%
Fence	\$9,000	$\frac{9,000}{441,600} = \frac{45}{2208}$	0.0204	2.04%
Total	\$441,600	$\frac{441,600}{441,600} = 1$	1.0	100%

Answers will vary.

Possible response:
Area of larger rhombus:
$(180 \times 240) \div 2 = 21{,}600$ *square yards*
Area of small(red) rhombus:
$(60 \times 80) \div 2 = 2{,}400$ *square yards*

Cost to pant the roses (inside the smaller rhombus):
$2{,}400 \times \$100.25 = \$240{,}600$
Area where grass will grow:
$21{,}600\ yd^2 - 2{,}400\ yd^2 = 19{,}200$ *square yards*
Cost to plant the grass:
$19{,}200 \times \$10 = \$192{,}000$

Cost to secure the perimeter which is the sum of the congruent sides of the larger rhombus. The side also represents the hypothenuse of the right triangle shown above.
Hypothenuse of right triangle:
$c^2 = a^2 + b^2$
$a = 90$ *yards or ½ smaller diagonal of the larger rhombus.*
$B = 120$ *yards or½ larger diagonal of the larger rhombus.*
$c^2 = 90^2 + 120^2$
$c^2 = 8{,}100 + 14{,}400 = 22{,}500$ *yards*

$c = \sqrt{22{,}500}$ $= 150$ *yards*

Perimeter of the larger rhombus:
$4 \times 150\ yd = 600$ *yards or1,800 feet.*
Cost to secure the perimeter:
$1{,}800 \times \$5 = \$9{,}000$
Total cost of the renovation:
$\$240{,}600 + \$19{,}200 + \$9{,}000 = \$441{,}600$
Answer: *The total cost of the park renovation is $441,600*

SOLO PRACTICE-25

Construct a similar problem. Solve it in multiple ways. Share it with the class.

Answers will vary.

GLOSSARY OF CLUSTERED & RELATED TERMS

Question	What is asked to find / Objective of a task
Step	One of the many phases of a procedure or solution process
Pathway	Set of steps leading to a solution
Process	Strategy and skills used to solve a problem
Solution Process	Pathway toward the solution of a problem
Skill	Ability to perform tactical operations / Procedural knowledge
Strategy	Grand plan toward a problem solution
Analyze	Study-Break down -Break down into composing parts
Comment	Feedback on a production / Opinion of a reviewer
Investigate	Study to discover the truth discover the truth
Verify	Prove a stated fact, statement, a conjecture
Justify	To prove the accuracy or validity of a statement
Evaluate	Judge / Measure / Gauge / Opine on then value
Perform	Execute a task
Convert	Translate into a different form
Expression	Whatever is expressed
Pattern	Predictable repetition of events
Formula	General rule / Algebraic equation showing relationships
Inequality	Imbalance between two entities
Pythagorean Theorem	Mathematical law stating: "in any right triangle the square of the hypotenuse equals the sum of the squares of the other two sides
Blueprint	Sketch or design
Venn Diagram	Graphic organizer that compares two or more entities
Table	Graphic organizer with rows and columns
Frayer Model	Graphic organizer showing the characteristics of a concept
Anticipatory Guide	A mechanism to elicit prior knowledge via pre-set statements.
Math Riddles	Statements that personify mathematical concepts
Semantic Map	Graphic organizer seeking connections with the central concept
Circle Graph	Graphic organizer where the entire circle is taken as 100%
Graphic Organizer	Representation of a situation via table, flow chart, bar graph, etc.
Difference	Result of a subtraction / What makes things not the same
Sum	Result of an addition / Collection of all given elements / Total
Product	Result of multiplication
Quotient	Result of division
Remainder	Leftover from division / What remains after an equal distribution
Factor	Quantity that is multiplied. Example 3 or 4 in $3 \times 4 = 12$
Multiple	Quantity derived from a multiplication of numbers (factors)
Element:	Distinct entity of a set
Intersection	Subset of elements common to many given set
Inclusion	Subset of elements within a given set / Incorporation: Symbol $\subset$
Union	Set compiling all other sets, wherein elements is not repeated

GLOSSARY OF CLUSTERED & RELATED MATH TERMS
(continued)

Time Zone	Area on earth with the same time
Military Time	Time used by military personnel. Example, 1300 hour for 1:00 pm
Pacific Time	Standard time within the limits of the Pacific Time Zone
Eastern Time	Standard time within the limits of the Eastern Time Zone
Central Time	Standard time within the limits of the Central Time Zone
Mountain Time	Standard time within the limits of the Mountain Time Zone
Minute	One-sixtieth of an hour (time) or of a degree (angle)
Second	One-sixtieth of a minute (time) or of a minute (angle)
Hour	One-twenty-fourth of a day
Day	One rotation of the earth
Week	Sequence of 7 days from Monday to Sunday
Month	Sequence of 30 or 31 days; except February 28; but 29 in leap year
Semester	Sequence of 6 months
Trimester	Sequence of 3 months
Mph	Number of miles covered in one hour
Unit	Standard of measurement
Yard	Unit of linear measurement in the Customary System: 1 yd= 3 ft ≈ 1 meter
Foot	Unit of linear measurement in the Customary System: 1 foot = 12 inches
Inch	Unit of linear measurement in the Customary System: 1 inch = 1/12 foot
Meter	Unit of linear measurement in the Metric System: 1 meter = 1/40,000,000 of the Circumference of the Earth
Centimeter	Unit of linear measurement in the Metric System: 1 cm = 1/100 meter
Kilometer	Unit of linear measurement in the Metric System: 1 km = 1,000 meters
Millimeter	Unit of linear measurement in the Metric System: 1 mm = 1/1000 mete
Decameter	Unit of linear measurement in the Metric System: 1dam = 10 meters
Hectometer	Unit of linear measurement in the Metric System: 1 hm = 100 meters
Square unit	Unit of area measurement of a plane figure: Generic term used when specific unit is unknown
Ounce	One sixteenth of a pound
Pound	Value of 16 ounces
Kilogram	Unit of weight or mass measurement in the Metric System 1 kg = 2.2 pounds

GLOSSARY OF CLUSTERED & RELATED MATH TERMS
(continued)

Fraction	A part of one whole. A subset of a group of things
Numerator	Number of parts considered from a total of equal parts
Denominator	Total number of equal parts after a equal grouping or separation
Ratio	Rapport between two entities – Comparison between two entities
Compare	Tell what is similar about distinct things; Also tell of the difference
Percent	Numeric value when comparing to 100
Decimal	Number that involves parts of a whole
Nearest	The closest to a given entity
Square Root	Number that is raised to the given second-degree power
Scientific Notation	Expression of a number using a decimal and a power of ten
Transformation	Repositioning of a geometric shape on a Cartesian Plane or Coordinate System
Translation	Transformation where a figure slides in a certain direction
Reflection	Transformation where a figure flips over a symmetrical line
Rotation	Transformation where a figure turns about a given point
Dilation	Transformation in size while preserving similarity
Data	Given facts, information
Average	Mean on a statistical data / Value for an equal distribution
Mean	Statistical element or average; Per capita value in equal distribution
Mode	Most frequent element of the data set; Most repeated element
Median	Element located in the middle of an ordered line of data elements
Range	Difference between highest number and lowest number of a data set
Pi (π) or 3.1416	Ratio of circumference to diameter of circle / Irrational number
Diameter	Special chord crossing the circle though the center Diameter = 2 radii
Radius	Line segment joining the center of the circle and a point on the circumference
Circumference	Perimeter of circle = $\pi d = 2\pi r$ / Line that makes entire circle
Chord	Any line segment joining two points on the circumference of circle
Arc	Piece of the circumference
Sector	Part of the area of the circle limited by two radii
Compounding	Method of computing balance of a loan based on periodic renewal of the principal
Principal	Amount that is deposited, loaned, or borrowed
Interest	Profit earned by the principal
Rate	Rapport between the principal and the interest expressed in percent
Probability	Chance for an event to happen.
Experimental Probability	Chance of outcomes based on experiments or trials
Theoretical Probability	Chance of outcomes based on mathematical or scientific calculations

GLOSSARY OF CLUSTERED & RELATED MATH TERMS
(continued)

Even	Natural number divisible by 2
Randomly	Blindly, by hazard; Without looking
With Replacement	Probability where the number picked is put back into the set
Without Replacement	Probability where the number picked is not put back into the set
Angle	Zone of a plane figure limited by two rays
Supplementary Angles	Pair of angles with a sum of 180 degrees
Complementary Angles	Pair of angles with a sum of 90 degrees
Adjacent Angles	Pair of angles that are next to each other, with a common side
Vertical Angles	Pair of angles of opposite regions of a common vertex when two Vertical angles are congruent
Congruent Angles	Angles of equal measures
Reflex Angle	Angle that measures more than 180 degrees
Alternate Interior Angles	Angles on opposite sides of the transversal but inside of the parallel lines. Alternate interior angles are congruent
Alternate Exterior Angles	Angles located on opposite sides of the transversal but outside of parallel lines. Alternate exterior angles are congruent
Corresponding Angles	Angles on the same side of the transversal, but one inside and the other outside. Corresponding angles are congruent
Concave Polygon	Polygon where one angle "caves" into the shape; diagonal crosses outside shape
Convex Polygon	Polygon where all diagonal cross inside the polygon
Regular Polygon	Polygon with all congruent sides or all congruent angles
Degree	Unit measure of angles / Extent
Line Segment	Part of a line with 2 endpoints
Point	The smallest figure in Geometry / Dot
Altitude	One dimension of parallelogram being perpendicular to the base
Perimeter	Borderline of a plane figure
Side	Line segment limiting a polygon
Distance	Gap / line segment between two locations
Diagonal	Line segment joining two non-consecutive vertices of a polygon
Hypotenuse	Side opposite to the 90-degree angle in a right triangle The longest side, generally "c", proving that $c^2 = a^2 + b^2$
Triangle	Polygon with three angles or three sides
Scalene Triangle	Triangle with no congruent sides
Obtuse Triangle	Triangle where one angle measures more than 90 degrees
Acute Triangle	Triangle where all angles measure less than 90 degrees
Right triangle	Triangle with a 90° angle
Width	One dimension of a rectangle
Length	One dimension of a rectangle
Square	Rectangle with 4 congruent sides; Rhombus with right angles

GLOSSARY OF CLUSTERED & RELATED MATH TERMS
(continued)

Parallelogram	Quadrilateral with 2 pairs of parallel sides
Quadrilateral	Polygon with sides
Rhombus	Parallelogram with 4 congruent sides
Trapezoid	Quadrilateral with only one pair of parallel sides (USA & Canada)
Trapezium	Quadrilateral with only one pair of parallel sides (England)
Polygon	Space figures limited by line segments
Hexagon	Polygon with 6 sides or 6 angles
Pentagon	Polygon with 5 sides or 5 angles
Heptagon	Polygon with 7 sides or 7 angles
Octagon	Polygon with 8 sides or 8 angles
Nonagon	Polygon with 9 sides or 9 sides
Decagon	Polygon with 10 sides or 10 angles
Dodecagon	Polygon with 12 sides or 12 angles
Shaded Region	Darken area of a figure
Area	Number of square units within a 2-dimensional, plane figure
Volume	Number of cubic units in a 3-dimensional, space figure
Prism	Space figure whose surfaces or faces are polygons
Faces	Polygons that set the limits of prisms and pyramids
Rectangular Prism	Prism limited by 6 rectangles. Example: shoebox
Cylinder	Space figure whose two circular bases. Example: Roll of tissue paper
Lateral Area	Surface other than the bases. Surface on the side of the space figure
Surface Area	Total area of all the faces of a space figure
Height	Altitude of a space/plane figure. Line segment perpendicular to the base
Base	Line segment perpendicular to the altitude of a plane figure
Base Area	Area of a space perpendicular to the altitude of a space figure
Dimensions	Basic measures of a figure. Ex: Length and width of a rectangle
Monomial	Algebraic expression with one term
Binomial	Algebraic expression with two terms
Polynomial	Algebraic expression with many terms
Trinomial	Algebraic expression with three terms
Distributive Property	Multiplication of each term of one factor by those the other factor
FOIL Method	Distributive property using two binomial factors by multiplying the **<u>F</u>**irst terms of the factors, then the **<u>O</u>**uter terms of the factors, then the **<u>I</u>**nner terms of the factors and the **<u>L</u>**ast terms of the factors
Half-dollar	America coin equivalent to 50 cents
Dime	American coin equivalent to one-tenth of a dollar
Quarter	Coin equivalent to one-fourth of a dollar
Nickel	American coin equivalent to one-twentieth of a dollar
Penny	American coin equivalent to one-hundredth of a dollar
Variable	Entity that can change in an algebraic expression or equation.

GLOSSARY OF CLUSTERED & RELATED MATH TERMS
(continued)

Equation	Equality between two things. The equal sign (=) must be present
Function	Relation where each domain element (x) is associated with only one range element (y)
Algebraic Solution	Process of resolving a problem in algebraic terms using variables
Geometric Solution	Process of resolving a problem using geometric shapes
Graph	Graphic organizer showing relationship (function) between two or more entities
X-Axis	Horizontal axis in a Cartesian plane
Y-Axis	Vertical axis in a Cartesian plane
Coordinate Plane	Cartesian plane or grid / Coordinate system identified by two perpendicular lines or axes XX' and YY", intersecting at a O, origin of the axes
Slope	Steepness of a line / Constant and ratio between the coordinates of the points on the graph of a linear equation or function
Y-Intercept	Point of intersection between a graph and the Y-Axis
X-Intercept	Point of intersection between a graph and the X-Axis
Origin	Beginning of the axes /Intersection of the axes, re: Cartesian plane

REFERENCES

Bayazit, Ibrahim. (2013). An investigation of problem-solving approaches, strategies, and models used by the 7th and 8th grade students when solving real-world problems. *Educational Sciences: Theory and Practice, v13n3p1920-1927 Sum 2013.8 pp*

Bonnesen, C. T. (2021). The nature of mathematics: Let's talk about it. Mathematics Teacher. 114(5):352-361.DOI:10.5951/MTLT.20200226.

Bryant & Hammill (2000). Characteristics behaviors of students with LD who have teacher-identified math weaknesses. SAGE.

Common Core State Standards (CCSS 2010). State standards initiative. http://www.corestandards.org

Danielson, C. (2007). *Enhancing professional practice: A framework for teaching, 2nd Edition.ASCD.* 1703 N. Beauregard St. Alexandria, VA> 22311-1714. USA.

Davis, J. D., Maeda, Y., Kahan, J., et al. (2008). Standards-based mathematics curricula and middle-grade students' performance on standardized achievement tests. *Journal for Research in Mathematics Education, 39*(3), 184-212.

Dewey, J. (1938). *Experience and education.* West Lafayette, IN: Kappa Delta Pi.

Douglas, E., Koro-Ljungberg, M., & McNeill, N. (2012, December). Moving beyond formulas and fixations: Solving open-ended engineering problems. *European Journal of Engineering Education, 37*(6), 627-651.

Gullie, K. (2011). Open-ended (extended/constructed response questions as predictors of success on subsequent State Mathematics Examinations: the influence of mathematical awareness and conceptual understanding. A Dissertation Submitted to the University at Albany, State University of New York in partial fulfillment of the requirements for the degree of Doctor of Philosophy.

Hiebert, J. (1997). Making sense: Teaching and learning mathematics with understanding. Heinemann, Portsmouth, NH.

Inoue, N., & Buczynski, S. (2011). You asked open-ended questions, now what? Understanding the nature of stumbling blocks in teaching inquiry lessons. *The Mathematics Education*, *20*(2), 10-22.

Joseph, A. (2020). A case study of mathematics teachers' perceptions of Haitian students solving constructed-response questions in New York middle schools. A Dissertation submitted to Concordia University Chicago in partial fulfillment of the requirements for the degree of Doctor of Education. Peer-reviewed. Published by ProQuest.

Johnson, D. and Johnson, R. (1998). Active learning: Cooperation in the college classroom. ERCIC Number: ED449714. ISBN-0-939603-14-4.

Leinwand, S. (2000). Sensible mathematics: A guide for school leaders. Heineman. ISBN-10.

Ma, L. (1999). Knowing and teaching elementary mathematics: teachers' understanding of fundamental mathematics in China and the United States. Mahwah, New Jersey: Lawrence Erlbaum Associates.

National Council of Teachers of Mathematics or NCTM. (2000). *Principles and standards for school mathematics*. NCTM.

National Research Council. (1989). *Everybody counts: A report to the nation on the future of mathematics education.* Washington, DC.

Noriyuki, I. (2011). Zen and the Art of Neriage: Facilitating Consensus Building in Mathematics Inquiry lesson. *Journal of Mathematics Teacher Education, 14(1),* p. 5–23. (February 2011).

Zen and the Art of Neriage: Facilitating Consensus Building in Mathematics Inquiry lesson through Lesson Study

O'Brien, T. C. & Moss, A. (2004). Real math? *The Phi Delta Kappan. 2004 / 12 Vol. 86; Iss. 4.*

Piaget, J. (1964). Development and learning. In R.E. Ripple Rockcastle (Ed.), *Piaget rediscovered* (pp. 7-20). Reprinted in *Readings on the development of children*. University of California, San Diego.

Program for International Student Assessment or PISA. (2015). *PISA results in focus: Better policies for better life*. OECD. 2016.

Schmoker, M. (2011). *Focus: Elevating the essentials to radically improve student learning.* ASCD. Alexandria, Va. 22311-1714 USA.

Singham, M. (2005). *The achievement gap in U.S. education: Canaries in the mine.* Lanham, MD. Rowman & Littlefield Education.

Thorndike, E. L. (1931). *Human learning.* New York, NY: The Century Company.

Trend International in Mathematics & Science Study or TIMSS. (2015). *TIMSS 2015 and TIMSS advanced international results in mathematics.* Boston, MA: TIMSS & Pirls International Study Center, Lynch School of Education, Boston College.

Vygotsky, L. S. (1978). *Interaction between learning and development.* Cambridge, MA: Howard Press.

Watson, J. B. (1930). *Behaviorism* (rev. ed., 6th printing, 1966) Chicago, IL: University of Chicago Press.

Wei, K. (May 2014). Explainer: What makes Chinese math lessons so good? http://the conversation.com/explainer-what-makes-chinese-maths-lessons-so-good-24380

About the Author

Archangelo Joseph, Ed.D., is a New York-based education consultant. He provides consultancy services to the New York City Department of Education and to the private firm Generation Ready, LLC. Along with his wife Julienne, he has founded Organic-way Mathematics Consulting & Publishing, LLC. For 30 years, he has served public school children in multiple capacities: Bilingual Teacher, District Staff Developer, NYC Headquarters (Tweed) Resource Specialist, Math Coach, Assistant Principal, and New York State (Language RBERN) Resource Specialist. Among all of them, he prides himself to be a prolific curriculum and instruction designer, particularly in K-8 school mathematics and English as a New Language (ENL), formerly English as a Second Language (ESL).

In 2020, Dr. Joseph received his degree of Doctor of Education within the School Leadership Program at Concordia University Chicago. Espousing the constructivist epistemology of Lev S. Vygotsky and Jean Piaget, the doctoral candidate had conducted a 7-year research "A Case Study of Mathematics Teachers' Perceptions of Haitian Students Solving Constructed-Response Questions in New York Middle School" examining why students can't solve problems requiring justifications of answers. Before then, Archangelo had graduated from the College of Saint Rose, Albany, New York, with two Advanced-Study degrees (School Supervision and District Supervision). Equally important, he had mesmerized classmates, chiefly Evelyn Gerdes, at City College of The City University of New York City College (CUNY) with a Master of Science degree in Education with a (4.0 GPA).

Besides leveraging school leadership and pedagogy skills respectively among New York school decision-makers and teachers, Dr. Joseph along with his spouse and a few friends/colleagues traveled to his native Haiti in 2017 to establish the "Centre de Ressources Éducatives du Bas-Artibonite" (Center for Educational Resources of Bas-Artibonite). Over there, the team captivated educational stakeholders (300 teachers and 100 school principals) in the City of Saint-Marc and the City of Gonaives by the wisdom regarding best practices in school pedagogy and in school supervision and administration.

Dr. Joseph, a tri-lingual (English, French, and Haitian Creole), lives happily in Nyack, New York, with his other half Julienne and their youngest daughter Shayenne who, now, is completing her freshman year within the medical program at Hampton University in Virginia.

Made in the USA
Monee, IL
02 November 2026

67384318R00046